PRINCIPLES *of* INFINITESIMAL STOCHASTIC *and* FINANCIAL ANALYSIS

PRINCIPLES *of* INFINITESIMAL STOCHASTIC *and* FINANCIAL ANALYSIS

IMME van den BERG
Universidade de Évora, Portugal

World Scientific
Singapore • New Jersey • London • Hong Kong

Published by

World Scientific Publishing Co. Pte. Ltd.

P O Box 128, Farrer Road, Singapore 912805

USA office: Suite 1B, 1060 Main Street, River Edge, NJ 07661

UK office: 57 Shelton Street, Covent Garden, London WC2H 9HE

British Library Cataloguing-in-Publication Data
A catalogue record for this book is available from the British Library.

ISBN 981-02-4358-8

Printed in Singapore.

Preface

It is my hope that it will make deep results from the modern theory of stochastic processes readily available to anyone who can add, multiply, and reason.
(Edward Nelson [16])

Aim, scope and method. The theory of options is considered to be one of the most efficient and applicable economic theories. It is based on a widely accepted, plausible and operational pricing model, originally formulated by Black and Scholes [4].

However, the spreading of this theory encounters some difficulties. This is essentially due to the heavy mathematical formalism associated to stochastic processes with continuous time, the basic tool of the Black-Scholes model [10], [14].

Our aim is to present the theory of option-pricing, while avoiding the continuous-time stochastic processes. Our starting point is the finite, time-discrete pricing model of Cox, Ross and Rubinstein (see [5]). This model is equally accepted and plausible, and consistent with the Black-Scholes model, but it is considered to be less operational.

Our approach will preserve both the efficiency of the first and the simplicity of the latter. Applying asymptotic methods on discrete formulae we derive in full generality the principal option-price formulae issued from the time-continuous model of financial markets.

This approach is made possible by using nonstandard analysis. At first sight, the reader might be discouraged by the use of nonstandard analysis. However, the nonstandard terminology is less used in a logical sense, to distinguish standard objects from nonstandard objects, than as a tool in asymptotics: it is a very convenient way to dispose of all kinds of orders of magnitude. A discourse in terms of orders of magnitude tends to be so elementary and intuitive that mathematical rigour does not consist an obstacle: often the discourse becomes close to very common heuristic arguments. Also, the existence of different orders of magnitude seems quite natural for someone observing the practice of financial markets: the amount of money involved is astronomical, trading is very fast, the price fluctuations usually are small, not excluding fluctuations of a higher order of magnitude, and occasional jumps.

The introduction of orders of magnitude provides a refinement of the Cox-Ross-Rubinstein model which makes it closer to reality; for instance, by requiring the trading-period to be infinitesimal we capture the practice of rapid price-adjustments. It makes the model also more operational, in particular because one is able to describe discrete systems with infinitesimal time steps using an efficient terminology inspired from continuous geometry. Indeed, time is denoted by t, the state of a random variable by x, we speak about S-continuous trajectories, discrete surfaces of trajectories .... In this context, the Cox-Ross-Rubinstein price of a European option (*i.e.* an option with a fixed exercise date) becomes a Riemann-sum. To derive the pricing formulae provided by the Black-Scholes theory is then straightforward: being integrals, they are simply the Riemann-integrals corresponding to the Riemann-sums. This very classical transition from the discrete to the continuous can be effectuated for large classes of options, again because of the conditions on the order of magnitude of the diverse parameters involved.

A rigorous approach to financial mathematics is generally reserved to post-graduate students. Due to the substantial simplification of the discrete setting a rigorous approach becomes accessible for a wider audience. This text is addressed to anyone with a background corresponding to a usual two-years course in econometrics or science. Indeed, the advance knowledge for this course is usually part of such a course: functions of two variables, parametric curves, Riemann-sums, improper Riemann-integrals, elementary probability, binomial coefficients, induction. The reader eager to know more about the mathematical implementation of the infinitely small and large is referred to introductory works like [8], [9], [18], [7], [12] and [17]. The last three works have effectively been teached at a first-year or second-year level. To my opinion, for non-mathematicians an investment in nonstandard analysis is more easy and rewarding than an investment in the mathematical background of continuous-time stochastics, which consists of high-level measure theory.

Summary: The course is divided into five chapters. Chapters 4 and 5 on stock prices and option pricing are preceded by three chapters presenting the pre-requisites on asymptotics, finite stochastic processes and their probability distributions. To be precise, we developed the tools which are necessary for an accurate discription of the self-financing hedging strategies and an effective calculation of stock and option prices.

In Chapter 1 we investigate the time-dependent binomial distribution and carry out a convenient re-scaling. In the new coordinates, which define a *binomial cone*, the time-dependent binomial distribution, now called *binomial function* becomes a sort of refined Pascal triangle, which is now infinitely close to the time-dependent Gaussian distribution: this constitutes a time-dependent De Moivre-Laplace theorem. This theorem, which in our context has become a theorem on uniform approximation on bounded domains, is proved in Chapter 2. The proof uses some nontandard tools. These tools are treated in the beginning of this chapter: nonstandard orders of magnitude, Riemann-sums with infinitesimal integration steps, principles of permanence of infinitesimal approximation and an asymptotic Chebyshev lemma, based on nonstandard notions of mass and tail of a random variable. The nonstandard De Moivre-Laplace theorem on the infinitesimal approximation of probabilities is followed by a theorem on infinitesimal approximation of expectations: under some general conditions expectations with respect to a binomial distribution are infinitely close to an expectation with respect to a normal distribution. This "continuization" theorem will be of primordial importance to derive the Black-Scholes option prices from the Cox-Ross-Rubinstein option prices.

In Chapter 3, we define finite stochastic processes, and distinguish among them the processes which are important for option-theory: bivalent processes, recombining processes, Markov processes, binomial processes, derivative processes, martingales and backward processes. We introduce a convenient geometric representation of recombining processes, in terms of discrete surfaces. Early in the chapter we define the discrete Wiener walk, which is central in the theory, and serves as illustration for many of the introduced notions and properties. When defining the crucial notions of conditional probability and conditional expectation, we avoided entirely algebra's of functions or sets, partitions and filtrations. Instead, we defined these notions for the two types of information which are usually available in financial markets: knowledge of the actual price, corresponding to what we called *state*-conditioning, and knowledge of the whole past behaviour of the price, which we called *trajectorial* conditioning. We show that the two forms of conditional expectation are confounded for a type of random variables which we call geometrically adapted (i.e. with respect to the geometry of the binomial cone, European options are of this type), and different for a type of random variables which we call trajectorially adapted (path-dependent options are of this type). We terminate the chapter with the martingale decomposition theorem and stochastic difference equations. These are discrete counterparts of subjects of the core of continuous-time stochastic processes, which in the finite setting become particularly elementary.

Not always did we introduce the above notions in full generality, and we avoided some notions of more traditional approaches. We did so because we preferred often concreteness and effectivity to abstraction. We remark that if at a latter stage the need is felt for more generality, our definitions are extended in a straightforward manner, and may serve as didactical particular cases. For instance, the *mass* of Chapter 1.3 is a concrete example of the notion of *almost surely* and trajectorial conditioning is a concrete example of finite filtration, the general setting of dealing with time-dependent information. Also, the Markov and martingale property which we defined in the geometric context of networks or trees are easily incorporated into such a setting.

In Chapter 4, we investigate in depth the discrete geometric Brownian motion and carefully motivate its choice as a model for the evolution of the stock prices. We try to do justice to the principal characteristics of the stock-market: rapid trading, unpredictable forecast even knowing the whole past behaviour, small fluctuations, rare jumps.

In Chapter 5, we define options as random variables. We distinguish path-dependent and path-independent (European) options. We describe the self-financing hedging strategy of Cox, Ross and Rubinstein in detail, using convenient numerical examples. The option price is the price of the initial hedging portfolio. We use the martingale property to show that the option price may be written as an expectation with respect to an artificial, "risk-neutral" probability. For path-dependent options the pricing formula becomes a sort of discrete path-integral. In the case of European options, we use the binomial function to reduce the pricing formula to a Riemann-sum, which by the methods of Chapter 2 is transformed into a Riemann-integral ("Feynman-Kač" formula). We consider this derivation to be one of the main features of this book and summarized it in section 5.3. Finally, the important Black-Scholes formula is obtained by applying this formula to call-options.

Numerous exercises illustrate the material of the main text. Most of them are oriented towards the explicit, or approximate calculation of probabilities, expectations, stock prices and option prices. Some of them concern Poisson distributions and jump-processes, complementing the processes of binomial type, which are the most common in this book.

Acknowledgements. This study is based on the work of Cutland, Kopp and Willinger, who first had the idea to use a nonstandard Cox-Ross-Rubinstein model for option pricing in [6] (this paper has been followed by several others), the introductory work on nonstandard discrete stochastic processes of E. Nelson [16], and on two previous publications of Koudjeti and myself ([3] and [13]).

The text is an extension of the notes, in French, of a course which I presented in the Spring of 1996 at the University of Nice to third-year students of MASS (Mathematics Applied to Social Sciences), at the invitation of F.Diener and M.Diener. In the first two years of MASS at the University of Nice, a large part of the courses is given in the context of nonstandard analysis. This fact, of course, facilitated the teaching of the present course. The text took a large profit from a second and third teaching experience by Mrs. and Mr. Diener in 1997 and 1998. I sincerely thank Francine and Marc Diener for giving me the opportunity to present this course and for participating, in a very active way and in a team spirit, in the elaboration of this text. I thank them, the students and all other colleagues who followed this course for their numerous suggestions and improvements.

I thank Theo Dijkstra of the SNS-Bank/ University of Groningen for guiding me through the world of mathematical finance and Hans Nieuwenhuis of the University of Groningen for almost dayly discussions on the subject, during a long period. Paulo Correia of the University of Évora provided very appreciated help with several computer problems.

I thank very much Fouad Koudjeti of the Kas-Associatie, Amsterdam for his translation of the original French notes ´´Introduction au calcul infinitésimal stochastique et financier´´ into English. This translation was the basis of the actual text.

Évora, 2000
Imme van den Berg

Contents

Chapter 1

The binomial cone and the binomial coefficients

In this chapter we present the framework of our theory. The finite stochastic processes will usually be defined on networks of points which we introduce in Section 1.1. In many cases their probability distributions will be binomial.

The binomial distribution will be recalled in Section 1.2, and then it will be adapted to the discrete network by a change of scale.

1.1 The binomial cone

Definition 1.1 Let $\delta t > 0$. The *arithmetic network* $\mathcal{R}\left(\delta t, \sqrt{\delta t}\right)$ is defined by

$$\mathcal{R}\left(\delta t, \sqrt{\delta t}\right) = \left\{(t, x) \in \mathbb{R}^2 \;\middle|\; \begin{array}{c} \exists \nu, j \in \mathbb{Z}, \\ t = \nu \delta t\,,\ x = (2j - \nu)\sqrt{\delta t} \end{array}\right\}.$$

The *binomial cone* $\mathcal{C}\left(\delta t, \sqrt{\delta t}\right) \subset \mathcal{R}\left(\delta t, \sqrt{\delta t}\right)$ is defined by

$$\mathcal{C}\left(\delta t, \sqrt{\delta t}\right) = \left\{(t, x) \in \mathcal{R}\left(\delta t, \sqrt{\delta t}\right) \;\middle|\; t \geq 0, |x| \leq t/\sqrt{\delta t}\right\}.$$

Also, we set $\mathbb{T} = \{\nu \delta t \mid \nu \in \mathbb{N}\}$, $\delta x = 2\sqrt{\delta t}$, and for any fixed $t \in \mathbb{T}$, we denote by $\mathcal{C}_t$ the vertical section

$$\mathcal{C}_t = \left\{x \;\middle|\; (t, x) \in \mathcal{C}\left(\delta t, \sqrt{\delta t}\right)\right\}, \tag{1.1}$$

1

and for $T \in \mathbb{T}$, the *binomial triangle* is the set

$$\mathcal{C}_{[0..T]} = \left\{ (t,x) \in \mathcal{C}\left(\delta t, \sqrt{\delta t}\right) \;\middle|\; t \le T \right\}, \tag{1.2}$$

where $[0..T]$ is the set $\{0, \delta t, 2\delta t, \dots, T\}$.

If $\delta t \simeq 0$, the above mentioned networks are referred to as *infinitesimal*.

The network $\mathcal{R}\left(\delta t, \sqrt{\delta t}\right)$ is a network of diamonds. To get a first idea, we will draw the network for $\delta t = \sqrt{\delta t} = 1$ (in this case, the diamonds are actually squares). The network is then

$$\mathcal{R}(1,1) = \left\{ (\nu, 2j - \nu) \in \mathbb{R}^2 \;\middle|\; \nu, j \in \mathbb{Z} \right\}.$$

Its vertical sections are alternatively odd and even integers. In this case, the cone $\mathcal{C}\left(\delta t, \sqrt{\delta t}\right)$ becomes

$$\mathcal{C}(1,1) = \{ (\nu, 2j - \nu) \mid \nu \in \mathbb{N}, 0 \le j \le \nu \}.$$

Its vertical sections $\mathcal{C}_\nu$ are collections of precisely $\nu + 1$ points (see figure 1.1).

Usually, δt wi ll be supposed to be infinitesimal, in which case $\sqrt{\delta t}$ is also infinitesimal but yet (infinitely) larger than δt. If there is no possible confusion on the period δt under consideration, we will write $\mathcal{C}$ instead of writing $\mathcal{C}\left(\delta t, \sqrt{\delta t}\right)$.

Sometimes, we will like to go back and forth between the discrete notation of the points of the network using integer coordinates and a "continuous" notation using real coordinates t and x. To do that, we adopt the following convention:

Notation : For $(t,x) \in \mathcal{R}\left(\delta t, \sqrt{\delta t}\right)$, we denote:

$$\begin{cases} \nu_t &= t/\delta t \\ j_{t,x} &= \frac{t}{2\delta t} + \frac{x}{2\sqrt{\delta t}} \end{cases} \tag{1.3}$$

Inversely, if $\nu, j \in \mathbb{Z}$, we write:

$$\begin{cases} t_\nu &= \nu \delta t \\ x_{\nu,j} &= \left(j - \frac{\nu}{2}\right) \cdot 2\sqrt{\delta t} \end{cases} \tag{1.4}$$

If $\nu \ge 0$ and $0 \le j \le \nu$, then $(t_\nu, x_{\nu,j}) \in \mathcal{C}\left(\delta t, \sqrt{\delta t}\right)$.

1.2 Rescaling

Definition 1.2 Let $\nu, j \in \mathbb{N}$, such that $0 \leq j \leq \nu$. Recall that

$$\binom{\nu}{j} = \frac{\nu!}{j!(\nu-j)!} \ (= C_\nu^j).$$

Let $0 < p < 1$. Then the *binomial coefficient* $B_p(\nu, j)$ is defined by

$$B_p(\nu, j) = \binom{\nu}{j} p^j (1-p)^{\nu-j}.$$

It is well-known that the binomial coefficients define a probability law on the set $\{0, 1, \dots, \nu\}$. Indeed, if $\mathrm{pr}\,(j) = B_p(\nu, j)$ then $\mathrm{pr}\,(j) \neq 0$ for all

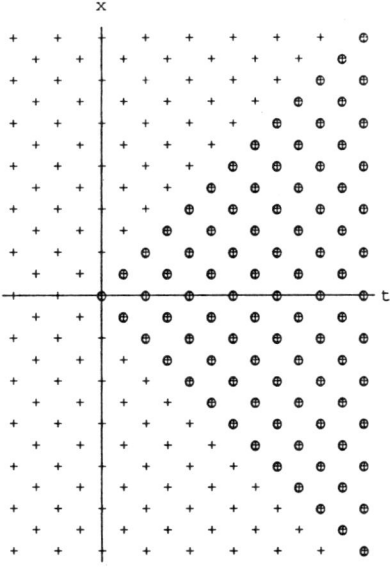

Fig. 1.1 The binomial cone $\mathcal{C} = \mathcal{C}\left(\delta t, \sqrt{\delta t}\right)$.

$j \in \{0, 1, \dots, \nu\}$, and we have

$$\sum_{j=0}^{\nu} \mathrm{pr}\,(j) = \sum_{j=0}^{\nu} B_p(\nu, j)$$

$$= \sum_{j=0}^{\nu} \binom{\nu}{j} p^j (1-p)^{\nu-j} = (p + (1-p))^\nu = 1^\nu = 1.$$

It is also well-known that if x is a random variable on a probability space taking its values in $\{0, 1, \dots, \nu\}$ and such that $\mathrm{pr}\,\{x = j\} = B_p(\nu, j)$ then its mean μ is equal to νp and its standard deviation σ is equal to $\sqrt{\nu p (1-p)}$.

Here is an example of such a random variable. Suppose that, similarly to a head-or-tail game, the outcomes of a random variable are either 1 with probability p or -1 with probability $1-p$. If the outcomes are independent then the probability that in ν tosses we get j times the outcome 1 is equal to $B_p(\nu, j)$. We will find this kind of situations in some finite stochastic processes, in particular for those processes defined on the binomial cone.

The binomial function we introduce here will prove to be a useful tool for studying such processes. It is the representation of the binomial coefficients seen at a new scale. Let $p := p_a := \frac{1}{2} + a\sqrt{\delta t}$, where $a \in \mathbb{R}$. We adopt this notation because the parameter p of most of the binomial coefficients B_p we will come across is near to $1/2$, and to be more precise, at a distance of the order of $\sqrt{\delta t}$ from $1/2$.

Definition 1.3   Let a be a real number such that $0 < p = \frac{1}{2} + a\sqrt{\delta t} < 1$. A *binomial function* is a function defined on the binomial cone $\mathcal{C}$ by

$$b_a(t, x) = \frac{1}{\delta x} B_{\frac{1}{2} + a\sqrt{\delta t}}(\nu_t, j_{t,x})$$

If $a = 0$, we will write $b(t, x)$ instead of $b_0(t, x)$.

If δx is infinitesimal, the binomial function $b_a(t, x)$ is the result of a change in scale of the family of binomial coefficients in four aspects:

- The mapping

$$B_{\frac{1}{2} + a\sqrt{\delta t}} \longrightarrow b_a$$

  is a *microscope* on the values of the binomial coefficients. It enlarges them with the factor $\frac{1}{\delta x}$.

- The mapping

$$\frac{t}{\delta t} \longrightarrow t$$

is a *macroscope* on the integers ν which reduces the time periods with a factor δt. We recall that a macroscope is a change of scale which, opposite to a microscope, shrinks what is large.

- The mapping

$$\frac{t}{2\delta t} + \frac{x}{2\sqrt{\delta t}} \longrightarrow x$$

is a *telescope* – that is mobile macroscope – on the integers j. This mapping centralizes the integers j around $\frac{\nu_t}{2}$ (the mean of a random variable with the law $B_{\frac{1}{2}}$), and it reduces the distance between two consecutive integers with a factor δx (which is of the order of $\frac{2}{\sqrt{\nu_t}}$, the inverse of the standard deviation of a random variable with the law $B_{\frac{1}{2}}$).

- The mapping

$$\frac{1}{2} + a\sqrt{\delta t} \longrightarrow a$$

is a *microscope* on the values of the parameter. It enlarges the difference between p_a and p with the factor $\frac{1}{\sqrt{\delta t}}$.

The changes of scale have the effect of making the set of binomial coefficients "visible". Indeed, the function $b_a(t, x)$ takes appreciable values for appreciable t and limited a and x. In fact we will see that the binomial function is infinitely close to the Gaussian function (see Figure 2.1 in the next chapter):

$$b_a(t, x) \simeq \frac{1}{\sqrt{2\pi t}} \exp\left(-\frac{(x - 2at)^2}{2t}\right)$$

The above-mentioned changes of scale are just chosen to ensure this closeness.

1.3 Exercises

Exercise 1.1   Let $\delta t \simeq 0$. Let $(\tau, \xi) \in \mathbb{R}^2$ be standard. Let us define
$[\tau] = \max\{t \in \mathbb{T} \mid t \leq \tau\}$ and $[\xi] = \max\left\{x \;\middle|\; x \in \mathcal{R}\left(\delta t, \sqrt{\delta t}\right), \; x \leq \xi\right\}$

(1) Why is $[\tau] \simeq t$ and $[\xi] \simeq \xi$?
(2) Put $\tau = 0$. Show that there exists $(t, x) \in \mathcal{C}\left(\delta t, \sqrt{\delta t}\right)$ such that $t \simeq 0$ and $x \simeq \xi$.
(3) Conclude that if $\tau > 0$, there exists $(t, x) \in \mathcal{C}\left(\delta t, \sqrt{\delta t}\right)$ such that $t \simeq \tau$ and $x \simeq \xi$.
(4) The limited part of the boundary of the cone $\mathcal{C}$ is infinitely close to a standard line. Which?

Exercise 1.2

(1) For $(t, x) \in \mathcal{C}\left(\delta t, \sqrt{\delta t}\right)$ verify that ν_t and $j_{t,x}$ are positive integers. To which integers correspond the points $(t, 0)$ of the horizontal axis?
(2) Suppose that $\delta t \simeq 0$. Show that the points $(t, x) \in \mathcal{C}\left(\delta t, \sqrt{\delta t}\right)$ for which ν_t and $j_{t,x}$ are standard are all infinitely close to $(0, 0)$.
(3) Give an explicit point $(t, x) \in \mathcal{C}\left(\delta t, \sqrt{\delta t}\right)$ such that t is limited and x is unlimited.

Exercise 1.3   (**Reminder about the binomial coefficients**) Let $0 < p < 1$, $N \in \mathbb{N}$. Let x be a random variable taking values in $\{0, 1, \ldots, N\}$ and such that

$$\mathrm{pr}\,\{x = j\} \;=\; \mathrm{B}_p(N, j).$$

Then

$$\mathrm{E}\,x = \sum_{1 \leq j \leq N} j \binom{N}{j} p^j (1 - p)^{N-j}$$

Show that $\mathrm{E}\,x = Np$. Also, show that $\mathrm{Var}\,x = \mathrm{E}\,x^2 - (\mathrm{E}\,x)^2$ may be written as

$$\mathrm{Var}\,x = \sum_{2 \leq j \leq N} j(j-1) \binom{N}{j} p^j (1 - p)^{N-j} - Np - (Np)^2,$$

and derive that $\mathrm{Var}\,x = Np(1 - p)$.

Exercise 1.4   Let $t \in T$. Let a be a real number such that $p := \frac{1}{2} + a\sqrt{\delta t}$ satisfies $0 < p < 1$.

(1) Check that $\left\{ b_a(t, x)\delta x \ \middle| \ |x| \leq t/\sqrt{\delta t} \right\}$ is a probability distribution on $\mathcal{C}_t$.

(2) Let y be a random variable taking values in $\mathcal{C}_t$ with the above probability distribution. Express the mean and variance of y in terms of the integers $j = j_{t,x}$ and $N = \nu_t$. Apply exercise 1.3 to show that the mean of y satisfies $E\, y = 2at$ and that the standard deviation of y satisfies $\sqrt{\operatorname{Var} y} = \sqrt{t(1 - 4a^2\delta t)}$.

(3) Suppose that δt is the inverse of an integer. For which values of a and t is the random variable y normalized (*i.e.* its mean is 0 and its standard deviation is 1)?

(4) Suppose that $\delta t \simeq 0$ and a is limited. For which values of t are the mean and the standard deviation of the random variable y limited? For which values are they infinitesimal?

Exercise 1.5   Consider the mapping

$$c: \quad \mathcal{C}\left(\delta t, \sqrt{\delta t}\right) \quad \longrightarrow \quad \mathbb{R}$$
$$(t, x) \quad \longrightarrow \quad \begin{pmatrix} j_{t,x} \\ \nu_t \end{pmatrix} \quad \left(= C_{\nu_t}^{j_{t,x}}\right).$$

Check that $c(t, x) + c(t, x + \delta x) = c\left(t + \delta t, x + \sqrt{\delta t}\right)$, *i.e.* the mapping c is a "Pascal triangle."

Chapter 2

Asymptotic properties of finite random variables

Probability distributions are very diverse. However, many of them are subject to unifying principles. Under quite general hypotheses, they obey the law of large numbers and/or are similar to the normal distribution of Gauss. A still larger unifying principle is the Chebyshev inequality. In this chapter, we will establish principles of that kind, which will prove useful for the study of binomial stochastic processes. Namely, we will show how the binomial distribution can be approximated by the normal distribution -a result known as the De Moivre-Laplace central limit theorem-. Then we will explain how to approximate the mean of a random variable with a binomial distribution by the mean of a corresponding random variable with the normal distribution.

Our tools are the nonstandard notions of *mass* and *tail* of a random variable and the *mass concentration lemma* deduced from the Chebyshev inequality.

We will also use more general properties like the approximation of an infinitesimal Riemann sum by a Riemann integral, and the fact that some approximative results remain valid beyond the domain where they are explicitly proved (as a consequence of *permanence principles*).

2.1 Preliminaries

2.1.1 *Orders of magnitude*

In classical asymptotics, there are symbols to denote orders of magnitude of functions, such as the Landau-notations $o()$ and $O()$. Similarly, nonstandard asymptotics has symbols to denote orders of magnitude of real numbers. We will first introduce the notion of real numbers of the same order of magnitude.

Definition 2.1   Let $a, b \in \mathbb{R}$. We will say that b is *of the order of* a if there is a limited real number λ such that $b = \lambda a$. We will say that a and b are *of the same order (of magnitude)* if both b is of the order of a and a if of the order of b. In this case, the quotients b/a and a/b are appreciable.

Notation :   A non-explicitly stated limited real number will be denoted by £, while a non-explicitly stated infinitesimal will be denoted by $\oslash$. A non-explicitly stated positive appreciable real number will be denoted by @. A non-explicitly stated positive unlimited number will be denoted by ∞. In general, neither do two occurrences of the symbol £ in a formula denote the same limited real number, nor do two occurrences of the symbol $\oslash$ denote the same appreciable number, nor do two occurrences of ∞ denote the same positive unlimited number.

Examples :

- $3 = £2$ and $2 = £3$. So 3 and 2 are of the same order ($2 = @3$).
- If $\varepsilon \simeq 0$ then $\varepsilon = £1$. But we do not have $1 = £\varepsilon$. So 1 is of the order of ε but the converse is not true.
- Let $\varepsilon \simeq 0$ and ω be a positive unlimited real number. Then $\varepsilon^2 = \oslash\varepsilon$, $\omega + 1 = (1 + \oslash)\omega$, $\sin \omega = £$, $\ln(1 + \varepsilon) = \varepsilon(1 + \oslash)$, $\exp £ = @$, $\ln @ = £$.
- Using these symbols, we can re-write the Leibnitz rules concerning the relations between arithmetic operations and orders of magnitude as follows:

$$
\begin{aligned}
£ + £ &= £, \\
£ \times £ &= £, \\
£ \times \oslash &= \oslash, \\
£/@ &= £, \\
\oslash + \oslash &= \oslash, \\
£/\infty &= \oslash, \\
&\vdots
\end{aligned}
$$

- Suppose that b is of the order of a. Then for any standard integer r, b^r is of the order of a^r. Indeed, $b^r = (£a)^r = £^r a^r = £a^r$.

2.1.2 *Riemann sums*

Definition 2.2   Let $\delta x > 0$. We will denote by $\mathbb{X}$ the set

$$\mathbb{X} = \{n\delta x \mid n \in \mathbb{Z}\}.$$

For $a, b \in \mathbb{X}$ such that $a \leq b$, the *discrete interval* $[a..b]$ is given by

$$[a..b] = \{x \in \mathbb{X} \mid a \leq x \leq b\}.$$

If $\delta x \simeq 0$ then the discrete interval will sometimes be called a *near-interval.*

Notation :   Let $a, b \in \mathbb{X}$ such that $a < b$ and let $f : [a..b] \longrightarrow \mathbb{R}$. We will mostly write $\sum_{a \leq x \leq b} f(x)\delta x$ instead of $\sum_{x \in [a..b]} f(x)\delta x$.

The following proposition is a consequence of the nonstandard definition of the Riemann-integral.

Proposition 2.3   *Let $a, b \in \mathbb{R}$ be limited, and let $f : [a, b] \longrightarrow \mathbb{R}$ be a standard Riemann-integrable function. We have*

$$\sum_{a \leq x < b} f(x)\delta x \simeq \int_a^b f(x)dx.$$

This proposition can be generalised to near-standard functions and thus allows us to calculate the standard part (also called *shadow*) of a Riemann-sum. This result, which is elementary, will play a primordial role in this course.

Proposition 2.4   (Shadow of a Riemann sum) *Let a and b be two limited real numbers in $\mathbb{X}$ such that $a < b$. Let $f_0 : \mathbb{R} \longrightarrow \mathbb{R}$ be a standard continuous function, and let $f : [a..b] \longrightarrow \mathbb{R}$ be a mapping such that $f(x) \simeq f_0(x)$ for all $x \in [a..b]$. Then*

$$\sum_{a \leq x < b} f(x)\delta x \simeq \int_a^b f_0(x)dx.$$

Proof : We have

$$
\begin{aligned}
\sum_{a \leq x \leq b} f(x)\delta x &= \sum_{a \leq x \leq b} f_0(x)\delta x + \sum_{a \leq x \leq b} (f(x) - f_0(x))\delta x \\
&= \int_a^b f_0(x)dx + \oslash + \sum_{a \leq x \leq b} \oslash \delta x \\
&= \int_a^b f_0(x)dx + \oslash(b - a + \delta x) \\
&= \int_a^b f_0(x)dx + \oslash.
\end{aligned}
$$

□

N.B. The function f_0 is usually called the *standard part* or *shadow* of f.

Contrary to the notion of shadow of a point, the notion of shadow of a function is rather intricate, and will be avoided here.

The following proposition is related to the well-known method for the majoration of a sum by an integral.

Proposition 2.5   *Let $a, b \in X$ with $a < b$. Let $f : [a, b] \longrightarrow \mathbb{R}$ be a decreasing, Riemann-integrable function. Then*

$$
\sum_{a < x \leq b} f(x)\delta x \leq \int_a^b f(x)dx.
$$

Proof : Since f is decreasing, we have that

$$
f(x)\delta x \leq \int_{x - \delta x}^x f(y)dy.
$$

Consequently,

$$
\sum_{a < x \leq b} f(x)\delta x \leq \sum_{a < x \leq b} \int_{x - \delta x}^x f(y)dy = \int_a^b f(y)dy.
$$

□

2.1.3   *Permanence principles*

In this course, we will use the following versions of two well-known permanence principles: *the Cauchy principle* and *Robinson's lemma.*

Lemma 2.6   (Cauchy principle) *Let f and g be two real-valued functions on $\mathbb{R}$ such that for all unlimited x we have $f(x) \leq g(x)$. Then there exists a limited real number A such that $f(x) \leq g(x)$ for all $x \geq A$.*

Lemma 2.7   (Robinson's lemma) *Let f and g be two real-valued functions defined on a (near-) interval D including limited and unlimited reals. If $f(x) \simeq g(x)$ for all limited $x \in D$ then there exists an unlimited real number $\alpha > 0$ such that for all $x \in D$ satisfying $|x| \leq \alpha$ we still have $f(x) \simeq g(x)$.*

Proof :   Let

$$\varepsilon_n = \sup_{-n \leq x \leq n, x \in D} |f(x) - g(x)|.$$

These numbers ε_n are infinitely small for all limited indices n. By the original version of Robinson's lemma, there exists an unlimited index α such that $\varepsilon_n \simeq 0$ for all $n \leq \alpha$. Hence, $f(x) \simeq g(x)$ for all $|x| \leq \alpha$.   □

2.2   Mass and tail of a random variable

A random variable can in principle take any value or set of values. Yet, it is very unlikely that a random variable takes a value which is very far from its mean, while it is probable that it takes values near to its mean. As a rule of thumb, we have *the statisticians confidence interval.* For example, we generally forecast that 95% of the results of some test lie in the interval $[\mu - 2\sigma, \mu + 2\sigma]$, where μ is the mean and σ the standard deviation of the sample. This is clearly an ad-hoc property and cannot be true in general, but it is well established for normally distributed samples.

The notion of *mass of a random variable* we are going to introduce is a kind of intrinsic confidence interval: it does not correspond to precise numerical values but to orders of magnitude. Similarly to the statistician's confidence interval, the mass of a random variable will be related to its mean and standard deviation. The *lemma of mass concentration* is simultaneously a Chebyshev inequality and an rudimentary central limit theorem. In fact, it will be useful in the proof of the latter.

Definition 2.8   Let (Ω, pr) be a probability space and $x : \Omega \longrightarrow \mathbb{R}$ be a random variable.

We will say that a real number λ is an element of the *tail* $\mathcal{Q}$ of x if either $\Pr\{x \geq \lambda\} \simeq 0$ or $\Pr\{x \leq \lambda\} \simeq 0$.   The *mass* $\mathcal{M}$ of a random variable x is the complementary in $\mathbb{R}$ of $\mathcal{Q}$. So, we have

$$\mathcal{M} = \bigcup_{\Pr\{x \leq \lambda_-\} \not\simeq 0 \text{ and } \Pr\{x \geq \lambda_+\} \not\simeq 0} [\lambda_-, \lambda_+].$$

Examples :

(1) Let x be a uniformly distributed random variable with values in $[2, 3]$. Then

$$\mathcal{Q} = \{x \mid x \simeq 2\} \cup \{x \mid x \simeq 3\}$$
$$\mathcal{M} = \{x \mid 2 \not\simeq x \not\simeq 3\}$$

(2) Let x be a random variable following a Cauchy law, with density function $f(x) = \frac{1}{\pi} \cdot \frac{1}{1+x^2}$. Then the mass of x is the collection of all limited reals and its tail is the collection of all unlimited reals. This can be shown directly by calculating the primitive of f (see exercise 2.12), or by the more theoretical argument just below.

(3) Consider a random variable such that its distribution law is normal, centred and reduced, *i.e.* its density function is $f(x) = \frac{1}{\sqrt{2\pi}} e^{-x^2/2}$. Since

$$\mathcal{N}(y) \equiv \frac{1}{\sqrt{2\pi}} \int_{-\infty}^{y} \exp\left(-\frac{x^2}{2}\right) dx$$

is a standard function strictly increasing from 0 to 1, one has firstly $\mathcal{N}(y) \simeq 0$ for all $y \simeq -\infty$, secondly $0 \not\simeq \mathcal{N}(y) \not\simeq 1$ for all limited y, and thirdly $\mathcal{N}(y) \simeq 1$ for all $y \simeq +\infty$. Again the mass of this random variable is the collection of all limited real numbers, and its tail is the collection of all unlimited reals. This will be the case of any normally distributed random variable with density of the form

$$f(x) = \frac{1}{\sqrt{2\pi}\sigma} \exp\left(-\frac{(x-\mu)^2}{2\sigma}\right)$$

where μ and $\sigma > 0$ are standard real numbers.

Next, we will present a theorem of localization of the mass of a random variable which is almost universal. We will see that the mass of a random variable is a domain which is centered around its mean, and whose size is of the order of magnitude of its standard deviation. This theorem is proved using the famous Chebyshev inequality.

Theorem 2.9 (Chebyshev inequality) *Let y be a positive random variable with finite mean. Then for all $\lambda > 0$,*

$$Pr\{y \geq \lambda\} \leq \frac{1}{\lambda} Ey.$$

Proof : We have

$$\mathrm{E}\, y \geq \mathrm{E}\, (y \cdot \chi_{\{y \geq \lambda\}}) \geq \mathrm{E}\, (\lambda \cdot \chi_{\{y \geq \lambda\}}) = \lambda \Pr\{y \geq \lambda\}.$$

□

Theorem 2.10 *Let y be a positive random variable with finite mean. Then each element of the mass of y is of order Ey.*

Proof : If $\mathrm{E}\, y = 0$, we have by the Chebyshev inequality that $\Pr\{y \geq \lambda\} = 0$ for all $\lambda > 0$. Since y is positive, we obtain that $\mathcal{M} = \{0\}$. Obviously, 0 is of order $\mathrm{E}\, y (= 0)$.

Suppose $\mathrm{E}\, y > 0$, and let λ be any unlimited real number. Using the Chebyshev inequality, we obtain

$$\Pr\{y \geq \lambda \mathrm{E}\, y\} \leq \frac{\mathrm{E}\, y}{\lambda \mathrm{E}\, y} \leq \frac{1}{\lambda} \simeq 0.$$

So $\lambda \mathrm{E}\, y$ is in the tail and hence, any element x of the mass satisfies $x = \lambda \mathrm{E}\, y$ for some limited λ.

□

Theorem 2.11 (Mass concentration lemma) *Let be given a random variable with finite mean μ and standard deviation σ. Then the distance between any point of its mass and μ is of order σ.*

Proof : Let x denote this random variable and put $y = (x - \mu)^2$. Let α be a point in the mass of $x - \mu$. Then $\Pr\{|x - \mu| \geq \alpha\} \gneq 0$ so that $\Pr\{y \geq \alpha^2\} \gneq 0$. By Theorem 2.10, α^2 is of order $\mathrm{E}\, y = \mathrm{Var}\, x = \sigma^2$. Hence α is of order σ.

□

Remark : Note that the mean and the standard deviation of random variables defined on finite probability spaces are always well-defined. In this context, the mass concentration lemma becomes a universal result.

2.3 Asymptotic properties of the binomial coefficients

The central-limit theorem says that, under quite general hypotheses, probability distributions are nearly normal. The De Moivre-Laplace theorem is its oldest form: it says that the binomial distributions B_p are nearly normal.

For the binomial processes we will use in option-theory, we will need the De Moivre-Laplace theorem for some values of p near to $1/2$, and to be more precise for those values of p of the form $p = \frac{1}{2} + a\sqrt{\delta t}$ where a is limited and $\delta t \simeq 0$. To establish the proof of the theorem for such values of p, we will need some appropriate approximations of the binomial coefficients. Roughly, our method is based on the following observation: to estimate the general term of a numerical sequence, say a_N, we may begin by estimating the first term a_0 and the quotients a_{i+1}/a_i, after which the estimation results from the fact that

$$a_N = a_0 \frac{a_1}{a_0} \frac{a_2}{a_1} \cdots \frac{a_N}{a_{N-1}}. \tag{2.1}$$

The proof of the form of the De Moivre-Laplace central limit theorem we are interested in will be achieved in three steps:

- the estimation of the quotient of two successive binomial coefficients, which we will establish in the Lemma's 2.12, 2.13 and 2.14. In the first lemma, we will obtain a combinatorial formula. The second and third lemma will restate this formula in the coordinates t and x, in an approximative way.
- an estimation of a binomial coefficient as a function of the maximal binomial coefficient, by successive approximations. This will be established in the first part of the proof of Theorem 2.15.
- the asymptotic approximation of the maximal binomial coefficient. We will be establish this in the last part of the proof of Theorem 2.15.

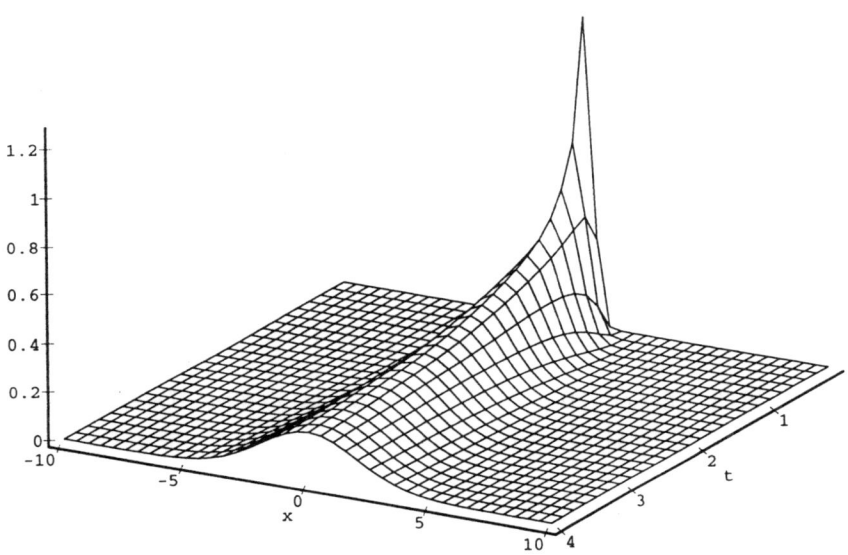

Fig. 2.1 The graph of the binomial function $b(x, t)$ is infinitely close to the graph of the Gauss function $\frac{1}{\sqrt{2\pi t}} \exp(-\frac{x^2}{2t})$.

Lemma 2.12 *Let* $0 < p < 1$, $j, \nu \in \mathbb{N}$, $0 \leq j \leq \nu$. *Then*

$$B_p(\nu, j+1) = B_p(\nu, j) \cdot \frac{1 - (j - \frac{\nu}{2})\frac{2}{\nu}}{1 + (j - \frac{\nu}{2})\frac{2}{\nu} + \frac{2}{\nu}} \cdot \frac{p}{p-1}.$$

Proof : We have

$$B_p(\nu, j+1) = \begin{pmatrix} \nu \\ j+1 \end{pmatrix} \cdot p^{(j+1)} \cdot (1-p)^{(\nu-j-1)}$$

$$= \frac{\nu!}{(j+1)!(\nu-(j+1))!} \cdot p^{(j+1)} \cdot (1-p)^{(\nu-j-1)} \cdot \frac{p}{1-p}$$

$$= \frac{\nu!}{j!(\nu-j)!} \cdot \frac{\nu-j}{j+1} \cdot p^j \cdot (1-p)^{(\nu-j)} \cdot \frac{p}{1-p}$$

$$= B_p(\nu, j) \cdot \frac{\frac{\nu}{2} - (j - \frac{\nu}{2})}{\frac{\nu}{2} + (j - \frac{\nu}{2}) + 1} \cdot \frac{p}{1-p}$$

$$= B_p(\nu, j) \cdot \frac{1 - (j - \frac{\nu}{2})\frac{2}{\nu}}{1 + (j - \frac{\nu}{2})\frac{2}{\nu} + \frac{2}{\nu}} \cdot \frac{p}{p-1}.$$

$\square$

Lemma 2.13 *Let $\delta t \simeq 0$ and a be a limited real number. Let t be appreciable. Then for any $(t, x) \in \mathcal{C}$,*

$$b_a(t, x + \delta x) = b_a(t, x) \cdot \frac{1 - \frac{x}{2t}\delta x}{1 + \frac{x}{2t}\delta x + \oslash \delta x} \cdot (1 + 2a\delta x + \oslash \delta x).$$

Proof : Put $\nu := \nu_t$, $j := j_{t,x}$, and $p := \frac{1}{2} + a\sqrt{\delta t}$. Note that

- $\dfrac{2}{\nu} = \dfrac{2}{t}\delta t = \dfrac{2}{t}\sqrt{\delta t} \cdot \sqrt{\delta t} = \oslash \delta x,$

- $(j - \dfrac{\nu}{2}) \cdot \dfrac{2}{\nu} = (j - \dfrac{\nu}{2}) \cdot \dfrac{2\delta t}{t} = \dfrac{(j - \frac{\nu}{2})\delta x}{2t} \cdot \delta x = \dfrac{x}{2t} \cdot \delta x,$

- $\dfrac{p}{1-p} = \dfrac{\frac{1}{2} + a\sqrt{\delta t}}{\frac{1}{2} - a\sqrt{\delta t}} = \dfrac{1 + a\delta x}{1 - a\delta x} =$
  $(1 + a\delta x)(1 + a\delta x + \oslash \delta x) = 1 + 2a\delta x + \oslash \delta x.$

Consequently,

$$b_a(t, x + \delta x) = \frac{B_p(\nu, j+1)}{\delta x}$$

$$= \frac{B_p(\nu, j)}{\delta x} \cdot \frac{1 - (j - \frac{\nu}{2}) \cdot \frac{2}{\nu}}{1 + (j - \frac{\nu}{2}) \cdot \frac{2}{\nu} + \frac{2}{\nu}} \cdot \frac{p}{1-p}$$

$$= b_a(t, x) \cdot \frac{1 - \frac{x}{2t}\delta x}{1 + \frac{x}{2t}\delta x + \oslash \delta x} \cdot (1 + 2a\delta x + \oslash \delta x).$$

$\square$

Lemma 2.14   *Let $\delta t \simeq 0$ and a be limited. Let $(t, x) \in C$ with appreciable t and limited x. Then*

$$b_a\left(t, x + \delta x\right) = b_a\left(t, x\right) \cdot \left(1 - \frac{x - 2at}{t}\delta x + \oslash \delta x\right)$$

Proof :   Since $\frac{x}{2t}\delta x \simeq 0$ we have $\frac{1}{1 + \frac{x}{2t}\delta x + \oslash \delta x} = 1 - \frac{x}{2t}\delta x + \oslash \delta x$. By Lemma 2.13 we get

$$
\begin{aligned}
b_a\left(t, x + \delta x\right) &= b_a\left(t, x\right) \cdot \frac{1 - \frac{x}{2t}\delta x}{1 + \frac{x}{2t}\delta x + \oslash \delta x} \cdot \left(1 + 2a\delta x + \oslash \delta x\right) \\
&= b_a\left(t, x\right) \cdot \left(1 - \frac{x}{2t}\delta x\right) \cdot \left(1 - \frac{x}{2t}\delta x + \oslash \delta x\right) \cdot \left(1 + 2a\delta x + \oslash \delta x\right) \\
&= b_a\left(t, x\right) \cdot \left(1 - \frac{x - 2at}{t}\delta x + \oslash \delta x\right).
\end{aligned}
$$

$\square$

Theorem 2.15   (De Moivre-Laplace) *Let $\delta t \simeq 0$ and a be limited. Then for any point $(t, x) \in C$ where t is appreciable and x limited,*

$$b_a\left(t, x\right) \simeq \frac{1}{\sqrt{2\pi t}} \cdot exp\left(-\frac{(x - 2at)^2}{2t}\right). \tag{2.2}$$

Proof :   We prove the theorem for the case where $a = 0$. The general case can be proven similarly by setting $x' \equiv -2at$.

For fixed t, the maximum of $b(t, x)$ is obtained for $x = 0$. We will first express $b(t, x)$ in terms of $b(t, 0)$, and then we will estimate $b(t, 0)$. We will use Lemma 2.14 and Proposition 2.4 which allow us to calculate the shadow of a Riemann sum.

For $x > 0$,

$$
\begin{aligned}
\mathrm{b}\,(t, x) &= \mathrm{b}\,(t, 0) \prod_{0 \le y < x} \frac{\mathrm{b}\,(t, y + \delta x)}{\mathrm{b}\,(t, y)} \\
&= \mathrm{b}\,(t, 0) \prod_{0 \le y < x} \left(1 - \frac{y + \oslash}{t}\delta x\right) \\
&= \mathrm{b}\,(t, 0) \cdot \exp\left[\sum_{0 \le y < x} \ln\left(1 - \frac{y + \oslash}{t}\delta x\right)\right] \\
&= \mathrm{b}\,(t, 0) \cdot \exp\left(\sum_{0 \le y < x} -\frac{y + \oslash}{t} \cdot \delta x(1 + \oslash)\right) \\
&= \mathrm{b}\,(t, 0) \cdot \exp\left(-\frac{1}{t}\sum_{0 \le y < x} (y + \oslash)\delta x\right) \\
&= \mathrm{b}\,(t, 0) \cdot \exp\left(-\frac{1}{t} \cdot \int_0^x y\, dy + \oslash\right) \\
&= \mathrm{b}\,(t, 0) \cdot \exp\left(-\frac{x^2}{2t} + \oslash\right) \\
&= \mathrm{b}\,(t, 0) \cdot \mathrm{e}^{-\frac{x^2}{2t}} \cdot (1 + \oslash).
\end{aligned}
\tag{2.3}
$$

The case where $x < 0$ can be treated in a similar way if one notes that

$$
\mathrm{b}\,(t, 0) = \mathrm{b}\,(t, x) \prod_{x \le y < 0} \frac{\mathrm{b}\,(t, y + \delta x)}{\mathrm{b}\,(t, y)}.
$$

To estimate $\mathrm{b}\,(t, 0)$, we will make use of the mass concentration lemma (Theorem 2.11) and Robinson's lemma (Lemma 2.7). Since $\mathrm{b}\,(t, x)\delta x$ is a probability distribution on C_t with mean 0 and standard deviation $\sqrt{t}$, the the mass concentration lemma ensures that for any unlimited z

$$
1 = \sum_{|x| \le \frac{t}{\sqrt{\delta t}}} \mathrm{b}\,(t, x)\delta x \simeq \sum_{x \le z} \mathrm{b}\,(t, x)\delta x.
$$

By Proposition 2.4 and Lemma 2.7, there exists some $z' \simeq +\infty$ such that

$$
\sum_{|x| \le z'} \frac{\mathrm{b}\,(t, x)}{\mathrm{b}\,(t, 0)}\delta x \simeq \int_{-z'}^{+z'} \exp\left(-x^2/2t\right) dx.
$$

Since the integral is convergent, we get

$$\int_{-z'}^{+z'} \exp\left(-x^2/2t\right) dx \simeq \int_{-\infty}^{+\infty} \exp\left(-x^2/2t\right) dx = \sqrt{2\pi t}.$$

We can now conclude that

$$1 \simeq \sum_{|x|\leq z'} b\left(t,x\right)\delta x = b\left(t,0\right) \cdot \sum_{|x|\leq z'} \frac{b\left(t,x\right)}{b\left(t,0\right)} \delta x$$
$$= b\left(t,0\right) \cdot \left(\sqrt{2\pi t} + \oslash\right).$$

Hence, $b\left(t,0\right) \simeq 1/\sqrt{2\pi t}$. Applying (2.3), we finally get

$$b\left(t,x\right) \simeq \frac{1}{\sqrt{2\pi t}} \cdot e^{-\frac{x^2}{2t}}.$$

$\square$

The De Moivre-Laplace theorem says that the binomial distribution behaves on its mass like the Gaussian function. This behaviour is lost on the tail, but only partly: the tail of the binomial distribution is still quadratically exponentially decreasing.

Proposition 2.16   *Let $\delta t \simeq 0$ and a be limited. For any $(t,x) \in C$, such that t is appreciable and x unlimited, we have*

$$b_a\left(t,x\right) = \pounds \cdot e^{-@x^2}.$$

Proof :   By Lemma 2.13, there exists some limited x_0 (say for example $x_0 = (4a+1)t$) such that for any $y \geq x_0$

$$b_a\left(t,y+\delta x\right) \leq b_a\left(t,y\right) \cdot \left(1 - \frac{y}{2t}\delta x\right).$$

Since for any $z < 1$ we have $\ln(1-z) \leq -z$, we deduce that for any unlimited

x

$$\mathrm{b}_a\left(t,x\right) \;\leq\; \mathrm{b}_a\left(t,x_0\right) \cdot \prod_{x_0<y<x}\left(1-\frac{y}{2t}\delta x\right)$$

$$= \;\mathrm{b}_a\left(t,x_0\right) \cdot \exp\left[\sum_{x_0<y<x}\ln\left(1-\frac{y}{2t}\delta x\right)\right]$$

$$\leq\; \mathrm{b}_a\left(t,x_0\right) \cdot \exp\left(\sum_{x_0<y<x}-\frac{y}{2t}\delta x\right)$$

$$\leq\; \mathrm{b}_a\left(t,x_0\right) \cdot \exp\left(\int_{x_0}^{x}-\frac{y}{2t}dy\right)$$

$$\leq\; \mathrm{b}_a\left(t,x_0\right) \cdot \exp\left(-\frac{x^2-x_0^2}{4t}\right)$$

$$= \;\pounds \cdot \mathrm{e}^{-@x^2}.$$

The inequality for $x < 0$ can be proved in a similar way. $\qquad\square$

The De Moivre-Laplace theorem states that binomial distributions are near to the corresponding normal distributions. It is then logical to wonder whether the expectation of a random variable with respect to a binomial distribution is near to the corresponding expectation with respect to the Normal distribution. Often, this is true indeed. When the binomial expectation is expressed in terms of the binomial function, it becomes a Riemann-sum. The expectation with respect to the normal distribution has the form of an integral, and under some regularity and growth conditions on the random variable this integral is the Riemann-integral corresponding to that Riemann-sum.

Because of the quadratically exponential decay of the binomial function and the Gaussian function, we expect convergence if the random variable has at most linear exponential growth. We say that a real function g defined for all $|x|$ large enough is of *exponential order at infinity* if there exist two constants K and C such that for all large enough $|x|$ one has $|g(x)| \leq Ke^{C|x|}$. If g is standard then, by transfer, the constants K and C can be chosen standard and the statement "for all $|x|$ large enough" can be replaced by "for all unlimited x." This inspires the notion of S-exponential order, which we define for any function, standard or not.

Definition 2.17 Let $D \subseteq \mathbb{R}$. A function $g : D \longrightarrow \mathbb{R}$ is said to be *of S-exponential order at infinity* if there are two standard reals K and C such that for any unlimited $x \in D$

$$|g(x)| \leq Ke^{C|x|}.$$

Note that if g is of S-exponential order at infinity then for any unlimited $x \in D$ we have

$$g(x) = \pounds e^{@x}.$$

Proposition 2.18 *Let $D \subseteq \mathbb{R}$ be a near-interval with limited and unlimited, and positive and negative elements. Let $g : D \longrightarrow \mathbb{R}$ be a function of S-exponential order which is infinitely close to a standard continuous function g_0 at each limited point x of D. Then g_0 is also of S-exponential order.*

Proof : Let K and C be standard and such that $|g(x)| \leq Ke^{Cx}$ for all unlimited $x \in D$. By the Cauchy principle, there exists a standard real number A such that $|g(x)| \leq Ke^{Cx}$ for all $x \in D$ which satisfy $|x| \geq A$.

Let y be any standard real such that $|y| \geq A$ and let $x \in D$ be such that $x \simeq y$. By hypothesis, we have

$$g_0(y) \simeq g_0(x) \simeq g(x) \leq Ke^{C|x|} \simeq Ke^{C|y|}.$$

So, $|g_0(y)| \leq Ke^{C|y|}$ by the Carnot principle*. By transfer, $|g_0(y)| \leq Ke^{C|y|}$ for any standard or nonstandard $|y| \geq A$, in particular for unlimited y. $\square$

Theorem 2.19 *Let $\delta t > 0$, $\delta t \simeq 0$. Let a be some limited real number and let $t \in \mathbb{T}$ be appreciable. Let C_t be the set given by (1.1) in Definition 1.1 provided with the binomial distribution B_a, and let $g : C_t \longrightarrow \mathbb{R}$ be a random variable of exponential order. Let g_0 be a standard continuous function, and suppose that for any limited x, $g(x)$ is infinitely close to $g_0(x)$. Then the expectation $E^{(a)}g$ of g satisfies*

$$E^{(a)}g \simeq \frac{1}{\sqrt{2\pi t}} \int_{-\infty}^{+\infty} \exp\left(-\frac{(x-2at)^2}{2t}\right) g_0(x)dx. \qquad (2.4)$$

---

*Carnot principle: any two standard reals which are infinitely close are necessarily equal.

Proof : We have

$$E^{(a)}g = \sum_{x=-t/\sqrt{\delta t}}^{x=+t/\sqrt{\delta t}} b_a\,(t,x)g(x)\delta x.$$

Let $y > 0$ be some limited real number. By Theorem 2.15 and Proposition 2.4 we get

$$\sum_{|x|\leq y} b_a\,(t,x)g(x)\delta x \simeq \frac{1}{\sqrt{2\pi t}} \int_{-y}^{+y} \exp\left(-\frac{(x-2at)^2}{2t}\right) g_0(x)dx.$$

By Robinson's lemma, there exists $\eta \simeq +\infty$ such that we still have

$$\sum_{|x|\leq \eta} b_a\,(t,x)g(x)\delta x \simeq \frac{1}{\sqrt{2\pi t}} \int_{-\eta}^{+\eta} \exp\left(-\frac{(x-2at)^2}{2t}\right) g_0(x)dx.$$

For $x \geq \eta$ we have the following majorations:

$$|b_a\,(t,x)g(x)| = \pounds\left(\exp -@x^2\right)\cdot e^{@x} = \pounds e^{-@x^2} \leq e^{-|x|},$$
$$\exp\left(-\frac{(x-2at)^2}{2t}\right) g_0(x) = \pounds\left(\exp -@x^2\right) e^{@x} \leq e^{-|x|},$$

so that by Proposition 2.5,

$$\left|\sum_{|x|>\eta} b_a\,(t,x)g(x)\delta x\right| \leq \sum_{|x|>\eta} e^{-x}\delta x \leq \int_{\eta}^{+\infty} e^{-x}dx = e^{-\eta} \simeq 0,$$
$$\left|\int_{|x|\geq\eta} \exp\left(-\frac{(x-2at)^2}{2t}\right) g_0(x)dx\right| \leq \int_{\eta}^{+\infty} e^{-x}dx = e^{-\eta} \simeq 0.$$

Hence,

$$\begin{aligned}
E^{(a)}g &\simeq \sum_{|x|\leq\eta} b_a\,(t,x)g(x)\delta x \\
&\simeq \frac{1}{\sqrt{2\pi t}} \int_{-\eta}^{+\eta} \exp\left(-\frac{(x-2at)^2}{2t}\right) g_0(x)dx \\
&\simeq \frac{1}{\sqrt{2\pi t}} \int_{-\infty}^{+\infty} \exp\left(-\frac{(x-2at)^2}{2t}\right) g_0(x)dx.
\end{aligned}$$

$\square$

2.4 Exercises

Exercise 2.1   Let $\omega \simeq +\infty$. Prove that the following real numbers are of different orders of magnitude:

$$\frac{1}{\omega} \; ; \; 1 \; ; \; \ln\omega \; ; \; \sqrt{\omega} \; ; \; \omega \; ; \; \omega^2 \; ; \; e^\omega \; ; \; \exp\left(\omega^2\right).$$

Exercise 2.2   Let $\varepsilon \simeq 0$.

(1) Consider the numbers $(k\varepsilon)^2$ for standard integers $k > 0$. Show that they are all of the same order of magnitude.
(2) Consider the numbers $exp(-k/\varepsilon)$, where k is a standard integer. Show that they are all of different orders of magnitude.
(3) Show that for standard integers $k > 0$ the numbers $\varepsilon^{1/k}$ are all of different order of magnitude. How about these values for nonstandard integers k?

Exercise 2.3   Give examples where the sum of an unlimited number of infinitesimals is infinitesimal, appreciable, and unlimited.

Exercise 2.4   Let $\varepsilon \simeq 0$ and $\eta \simeq 0$. Show that

$$\frac{1}{1 + \varepsilon(1 + \eta)} = 1 - \varepsilon(1 + \oslash)$$
$$\ln(1 + \varepsilon(1 + \eta)) = \varepsilon(1 + \oslash)$$
$$\exp(\varepsilon(1 - \eta)) = 1 + \varepsilon(1 + \oslash).$$

Exercise 2.5   Let $\delta x > 0$ and $\delta x \simeq 0$. Let $\mathbb{X}$ be the set of multiples of δx. We aim to show how different occurrences of the symbol $\oslash$ may represent different infinitesimals $\delta_1, \delta_2, \delta_3 \cdots$.

(1) Show that (trivially) the maximum and the minimum of a finite set of infinitesimals is infinitesimal.
(2) Show that

$$\sum_{0 \leq y < x, x \in \mathbb{X}} (1 + \oslash)\delta x = (1 + \oslash)x$$

(3) If x is limited and t is appreciable, show that

$$\sum_{0 \leq y < x, y \in \mathbb{X}} \ln\left(1 - \frac{y + \oslash}{t}\delta x\right) = \sum_{0 \leq y < x, y \in \mathbb{X}} -\frac{y + \oslash}{t}\delta x. \qquad (2.5)$$

Exercise 2.6   Let $\omega \simeq +\infty$. Show that

(1) $@\omega = \oslash\omega^2$,
(2) $@ \cdot \infty = \infty$
(3) $@\omega^2 + £\omega = @\omega^2$,
(4) $\exp\left(-@\omega^2 + £\omega\right) = \oslash$.
(5) $@^{£/\omega} = 1 + £/\omega$

Exercise 2.7   Investigate the possible values of $\oslash^\oslash$.

Exercise 2.8   Let $\delta t \simeq 0$, $\delta t > 0$, $\mathbb{T} = \{n\delta t \mid n \in \mathbb{Z}\}$, $\varepsilon \simeq 0$. We let always t be an element of $\mathbb{T}$. Give an infinitesimal approximation of the following quantities

(1) $\Sigma_{0\leq t<1} \ t^2\delta t$
(2) $\Sigma_{0\leq t<1} \ t^2(1 + \varepsilon\sin(e^t))\delta t$
(3) $\Sigma_{0\leq t<1} \ (1 + t\delta t)^{1/\delta t}\delta t$
(4) Let $\varepsilon(t) \simeq 0$ for $t \in [0,1[$. Use formula 2.5 to determine the standard part of

$$\sum_{0\leq t<1, t\in\mathbb{T}} \ln(1 - t(1 + \varepsilon(t))\delta t).$$

Exercise 2.9   Let $\alpha > 0$ be a limited but not necessarily standard real number.

(1) Show that $\Sigma_{1\leq t<2} \ \frac{1}{t+\alpha}\delta t \simeq \ln\left(\frac{2+\alpha}{1+\alpha}\right)$.
(2) Let $f : [0,1] \longrightarrow \mathbb{R}$ be a standard continuous function. Show that $\Sigma_{0\leq t<1} \ f(\alpha t)\delta t \simeq \int_0^1 f(\alpha t)dt$

Exercise 2.10   Let $\delta t \simeq 0$, $\delta t > 0$, $\mathbb{T} = \{n\delta t \mid n \in \mathbb{Z}\}$. We suppose that $t \in \mathbb{T}$. Using permanence principles, give an approximation of each of the following quantities

(1) $\Sigma_{1\leq t<\infty} \ \frac{1}{t^2} \cdot \delta t$
(2) $\Sigma_{-\infty<t<+\infty} \ \exp\left(-\frac{t^2}{2}\right)\delta t$
(3) $\Sigma_{0\leq t<\frac{1}{\delta t}} \ (1 - t\delta t)^{\frac{1}{\delta t}}\delta t$

Exercise 2.11   Let $\varepsilon > 0$. Let X_ε be a random variable with density function

$$f_\varepsilon(x) = \frac{1}{2\varepsilon}\exp\left(-\frac{|x|}{\varepsilon}\right).$$

(1) Determine the mean and variance of X_ε

(2) determine the mass of X_ε. For which ε does the mass contain only infinitesimals?

Exercise 2.12   Let V_α be a random variable with the Cauchy density function:

$$f_\alpha(x) = \frac{1}{\pi} \cdot \frac{\alpha}{\alpha^2 + x^2}.$$

(1) Let $\alpha \neq 0$ be standard. Estimate the probabilities $\Pr\{V_\alpha \le y\}$ and $\Pr\{V_\alpha \ge y\}$, and prove that the mass of V_α is the collection of all limited real numbers.

(2) What is the mass of V_α if $\alpha \simeq 0$?

Exercise 2.13   Let μ be limited and $\sigma \simeq +\infty$. Show that the mass of the normally distributed random variable $\mathcal{N}_{\mu,\sigma}$ is the collection of all reals of order σ.

Exercise 2.14   Let f be the density function of a random variable whose mass is included in the infinitesimals and let $\varepsilon_n = \int_{|x| \ge 1/n} f(x) dx$.

(1) Show that for any limited integer n we have $\varepsilon_n \simeq 0$.

(2) Using Robinson's lemma, conclude that there exists $\alpha \simeq 0$ such that the mass corresponding to the density f is contained in the interval $[-\alpha, \alpha]$.

(3) Let $y : \mathbb{R} \longrightarrow \mathbb{R}$ be a second random variable defined by

$$y(x) = \begin{cases} 0 & \text{if } |x| \ge 2 \\ 3x + 4 & \text{if } |x| < 2 \end{cases}$$

Show that $\mathrm{E}\,y \equiv \int_{-\infty}^{+\infty} y(x)f(x)\,dx \simeq y(0)(= 4)$.

Exercise 2.15   Let $(t, x) \in \mathcal{C}$ and $a \in \mathbb{R}$. Put

$$\begin{aligned} \beta(x) &= b_a(t, x) \\ \delta\beta(x) &= \beta(x + \delta x) - \beta(x). \end{aligned}$$

Let $\delta t \simeq 0$, a and x be limited and t be appreciable. Show from the results in the main text that β satisfies the approximate difference equation

$$\begin{cases} \delta\beta(x) & \simeq -(x - 2at)\beta(x) \\ \beta(0) & \simeq \frac{1}{\sqrt{2\pi t}} \end{cases} \tag{2.6}$$

Show also that $\beta(x)$ is infinitely close to the solution of the differential
equation

$$\begin{cases} G(x) & = -(x - 2at)G(x) \\ G(0) & = \frac{1}{\sqrt{2\pi t}} \end{cases} \qquad (2.7)$$

N.B. An alternative proof of the De Moivre-Laplace formula is based on
the direct transition of the difference equation (2.6) into the differential
equation (2.7). The general nonstandard method which justifies such a
transition from the discrete to the continuous is called *stroboscopy*. It falls
outside the scope of this course. In a sense our proof of the De Moivre-
Laplace theorem is based on a transition from the *solution* of (2.6) into the
solution of (2.7). Anyhow, in order to justify that the initial conditions are
(nearly) equal to $\frac{1}{\sqrt{2\pi t}}$, one can not do without a careful analysis of the
tails of binomial distribution, as we have done in the main text.

Exercise 2.16   Consider the Poisson distribution

$$\text{pr}_\lambda\{x = n\} = \frac{e^{-\lambda}\lambda^n}{n!} \qquad (n \in \mathbb{N}, \lambda > 0).$$

(1) Show that $\mathrm{E}\,x = \mathrm{Var}\,x = \lambda$.
(2) Show that if λ is limited the mass of x is contained in the collection
of limited real numbers, and that if λ is unlimited then its mass
is contained in the collection of unlimited reals. In the latter case,
give an example of an unlimited real number which is less than all
elements of the mass of x.

Exercise 2.17   Poisson distributions with large parameters are also nearly
normal. This is a result which is proven in a way similar to, but easier
than, that of the De Moivre-Laplace theorem. Let $\lambda \in \mathbb{N}$, $\lambda \simeq +\infty$. Let
Y be an infinite set of equidistant points of the form

$$\mathbb{Y} = \{(n - \lambda)\delta y \mid n \in \mathbb{N}\} \qquad \delta y = 1/\sqrt{\lambda}$$

and let x be a random variable taking its values in $\mathbb{Y}$. Assume its probabilty
distribution is given by

$$\text{pr}\{x = (n - \lambda)\delta y\} = \frac{e^{-\lambda}\lambda^n}{n!}.$$

For $y \in \mathbb{Y}$, put

$$r(y) = \frac{\mathrm{pr}\,\{x = y + \delta y\}}{\mathrm{pr}\,\{x = y\}}$$

(1) Prove that $\mathbb{Y}$ is chosen in such a way that x is normalized, i.e.
$\mathrm{E}\,x = 0$ and $\mathrm{Var}\,x = 1$

(2) Why is $\mathrm{Min}\,\mathbb{Y}$ negative unlimited?

(3) If $y = (n - \lambda)\delta y$, show that

$$r(y) = \frac{\lambda}{n+1}.$$

Express this formula in terms of y and δy, and for limited y, find the estimation

$$r(y) = 1 - y\delta y(1 + \oslash).$$

(4) Show that $\mathrm{pr}\,\{x = 0\}$ is maximal. Using Stirling's formula

$$\lambda! = (1 + \oslash)\lambda^{\lambda}e^{-\lambda}\sqrt{2\pi\lambda}$$

prove that

$$\mathrm{pr}\,\{x = 0\} = \frac{1 + \oslash}{\sqrt{2\pi}}\delta y.$$

(5) Deduce that for limited y we have

$$\mathrm{pr}\,\{x = y\} = \frac{1 + \oslash}{\sqrt{2\pi}}e^{-y^2/2} \cdot \delta y.$$

Consider separately the cases where y is positive and where y is negative.

N.B. It is possible to prove that $\mathrm{pr}\,\{x = 0\} = \frac{1 + \oslash}{\sqrt{2\pi}}\delta y$ in a way similar to the second part of the proof of De Moivre-Laplace theorem. This would have provided us with a proof of Stirling's formula. The possibility to prove Stirling's formula using Poisson distributions was already remarked by Emile Borel.

Exercise 2.18 Consider the binomial coefficients $B_p(\nu, j)$. Suppose that $\nu \simeq +\infty$ and $p = \lambda/\nu$, where $\lambda \geq 0$ is limited. Let ψ_λ be the Poisson-distribution:

$$\psi_\lambda(j) = \frac{e^{-\lambda}\lambda^j}{j!}$$

(1) Show that for standard j

$$B_{\lambda/\nu}(\nu, j) = \frac{\nu(\nu-1)\ldots(\nu-j+1)}{\nu^j} \cdot \frac{\lambda^j}{j!} \cdot \left(1 - \frac{\lambda}{\nu}\right)^{\nu-j} \simeq \psi_\lambda(j).$$

(2) Why is $B_{\lambda/\nu}(\nu, j) \leq \lambda^j/j!$ for all $j \leq \nu$?
(3) Let $j \simeq \infty$. Show that both $\psi_\lambda(j)$ and $B_{\lambda/\nu}(\nu, j)$ are superexponentially small, i.e. there exists $\epsilon \simeq 0$ such that they are less then ϵ^j.

Exercise 2.19 Let $t \in \mathbb{T}$ be appreciable, and let $g : \mathcal{C}_{[0..T]} \longrightarrow \mathbb{R}$ be a random variable provided with the binomial distribution b (T, x).

(1) Give an estimation of the expectation of g for $g(x) := e^{\left(x-\frac{T}{2}\right)}$ in terms of the Standard Normal distribution.
(2) Express the expectation of g in terms of Standard Normal distributions for

$$g(x) = \begin{cases} e^{\left(x-\frac{T}{2}\right)} - 1 & \text{if } x \geq T/2 \\ 0 & \text{if } x \leq T/2. \end{cases}$$

N.B. This last expectation represents the Black-Scholes formula for the value of a call-option in a very simplified context, but still containing its most essential elements (see Theorem 5.10 for $\xi_0 = K = 1$, $r = 0$ and $\sigma = 1$).

Chapter 3

Finite stochastic processes

The basic finite, discrete stochastic process is the Wiener walk, and the finite, discrete stochastic process which is fundamental in the theory of financial markets is the discrete geometric Brownian motion. In this chapter, we will present general notions and properties which will help us to deal with such processes. We will make use of the fact that both the above mentioned processes are *bivalent*, i.e at any moment there are two ways to continue, either by an upward movement or by a downward movement, and *recombining*, i.e. an upward movement followed by a downward movement give the same result as a downward movement followed by an upward movement. Guided by these two properties we give a presentation of stochastic processes which is more concrete than the usual presentations. The central notions will be the *trajectory*, modelling the (possible) development in time of uncertain quantities, and the *state*, determining the present situation of the trajectory. A particular class of processes may be visualized as discrete surfaces, i.e. as mappings defined on the binomial cone. This geometrical point of view facillitates the actual calculation of probability distributions and expectations.

3.1 Process and trajectories

Definition 3.1   Let $a, b \in \mathbb{T}$ such that $a \leq b$. A *trajectory* λ is a mapping from $[a..b]$ to $\mathbb{R}$. The *increment* $\delta\lambda$ of a trajectory is defined by $\delta\lambda(t) = \lambda(t + \delta t) - \lambda(t)$.

Definition 3.2 Let $[a..b] \subset \mathbb{T}$. A *finite stochastic process* η, indexed by $[a..b]$, is a finite set of trajectories, each of which is defined on $[a..b]$, provided with a probability pr .

For each t, we define a random variable η_t by $\eta_t(\lambda) = \lambda(t)$.

For $t \in \mathbb{T}$, $a \leq t < b$, the *increment* $\delta\eta_t$ of the process at t is defined by the random variable $\delta\eta_t = \eta_{t+\delta t} - \eta_t$. If λ is a trajectory and $t \in [a \ldots b]$ we denote the set of all trajectories which coincide with λ up to time t by $\overline{\lambda}_t$, i.e.

$$\overline{\lambda}_t = \{\alpha \in \eta \mid \forall s \leq t, \ \alpha(s) = \lambda(s)\}. \tag{3.1}$$

A finite probability space is a set of possible events, each of which has a weight. Similarly, a finite stochastic process is a set of possible future developments or scenario's on one and the same probability space (Ω, pr), each scenario being provided with a given weight or probability of occurrence. Hence, there are two ways of seeing the stochastic process η:

(1) **vertically:** by considering each one of its vertical sections η_t as random variables on the same probability space (Ω, pr). In this case, the process η is a collection of random variables defined on Ω. The index set is then $[a..b]$.

(2) **horizontally:** by considering the set Ω of possible scenario's in time, each scenario having its probability.

Example : Let $[a..b] = \{0, \delta t, 2\delta t, 3\delta t\}$ and let β be the process indexed on $[a..b]$ given in the following table

	0	1	2	3
$\beta^{(8)}$	0	1	2	3
$\beta^{(7)}$	0	1	2	1
$\beta^{(6)}$	0	1	0	1
$\beta^{(5)}$	0	1	0	-1
$\beta^{(4)}$	0	-1	0	1
$\beta^{(3)}$	0	-1	0	-1
$\beta^{(2)}$	0	-1	-2	-1
$\beta^{(1)}$	0	-1	-2	-3
	0	δt	$2\delta t$	$3\delta t$

Each line of the table represents one the eight trajectories $\beta^{(1)}$, $\beta^{(2)}$, , $\beta^{(8)}$ of the process β, and for $1 \leq j \leq 8$ we let pr $\beta^{(j)} = \frac{1}{8}$ (*i.e.* each one of the events $\beta^{(j)}$ occurs with a probability 1/8).

Here, we read the table line by line. We could also read it column by column: if $\Omega = \left\{ \beta^{(j)} \;\middle|\; 1 \le j \le 8 \right\}$, the vertical sections β_0, $\beta_{\delta t}$, $\beta_{2\delta t}$, $\beta_{3\delta t}$ are random variables on Ω with, for example, the following probabilities

$$\Pr\{\beta_{2\delta t} = -2\} = \frac{1}{4} \; , \quad \Pr\{\beta_{2\delta t} = 0\} = \frac{1}{2} \; , \quad \Pr\{\beta_{2\delta t} = 2\} = \frac{1}{4}$$

Ω

0	1	2	3
0	1	2	1
0	1	0	1
0	1	0	-1
0	-1	0	1
0	-1	0	-1
0	-1	-2	-1
0	-1	-2	-3
β_0	$\beta_{\delta t}$	$\beta_{2\delta t}$	$\beta_{3\delta t}$

with $\operatorname{pr}(\omega) = \frac{1}{8}$ for $\omega \in \Omega$.

The following table presents the increments $\delta\beta_0$, $\delta\beta_{\delta t}$, $\delta\beta_{2\delta t}$ of the process β under consideration

Ω

1	1	1
1	1	-1
1	-1	1
1	-1	-1
-1	1	1
-1	1	-1
-1	-1	1
-1	-1	-1
$\delta\beta_0$	$\delta\beta_{\delta t}$	$\delta\beta_{2\delta t}$

with $\operatorname{pr}(\omega) = \frac{1}{8}$ for $\omega \in \Omega$.

A practical and very frequent way of defining a stochastic process is by giving its initial value and the successive increments; this approach is analogous to the definition of a sequence of numbers by giving the initial value and the successive differences.

Indeed, we may determine the random variables η_t at any time $t \le b$ if we know the initial value of the process η_a (or otherwise stated, its first random variable), and the collection of the increments of the process in time until the moment prior to t, i.e. the collection of random variables $\{\delta\eta_s \mid a \leq s < t\}$. In fact, we take the sum of the initial value and the increments for $a \leq s < t$, and obtain $\eta_t = \eta_a + \sum_{a \leq s < t} \delta\eta_s$. As an exemple, one verifies that the random variables of the process β above are partial sums of its increments.

To determine the probabilities of the trajectories we consider first the case of two players throwing a coin: if it is tail then one of the players wins 1\$ (the asset of the first player increases with 1\$), but if it is head then it is the second player who gets 1\$ (which means a decrease of 1\$ in the asset of the first player). We will suppose that the probabilities of head and tail are both equal to $\frac{1}{2}$ and that the outcomes of successive tosses are mutually independent. Then a trajectory describing the gain of the first player is a function of the time, its end-value is equal to the *sum of increments* by 1 and -1, and the probability of the trajectory is equal to the *product of the probabilities of these increments*. In our example, it would precisely be a power of $(\frac{1}{2})$.

So the probability of a trajectory may be calculated from the probabilty of the increments if the latter are independent. In a general context, independence is defined as follows.

Definition 3.3   Let η be a finite stochastic process indexed by an interval $[a..b] \subset \mathbb{T}$. We say that the random variables $\{\eta_t \mid a \leq t \leq b\}$ are *independent* if for any trajectory λ of η we have

$$\text{pr}\,\lambda = \prod_{a \leq t \leq b} \Pr\{\eta_t = \lambda(t)\}. \tag{3.2}$$

If the increments a stochastic process η are independent in general the random variables itself are not independent. See exercise 1.

Saying that a random variable is independent from a family of random variables can be expressed in a more intrinsic way by saying that knowing the distributions of each random variable from that family does not tell us anything about the distribution of the random variable we are interested in. This way of expressing independence will be considered in Exercise 8.

We prove know some useful properties of the expectations and variances of independent random variables. Let x, y be two random varables defined on the same finite probability space Ω. We may identify the couple (x, y) with a stochastic process with one time-step. Then the trajectories correspond to the successive values $(x(\omega), y(\omega))$, assumed for the same element $\omega \in \Omega$. Within this interpretation, two random variables are independent if for all possible values α of x and β of y

$$Pr\{x = \alpha, y = \beta\} = Pr\{x = \alpha\}.Pr\{y = \alpha\}.$$

Then we have the following proposition.

Proposition 3.4 *Let x, y be two independent random variables defined on a same finite probability space. Then*

$$Exy = Ex.Ey$$

Proof : We have

$$
\begin{aligned}
E\,xy &= \sum_{\alpha \in \mathbb{R}} \alpha.Pr\{xy = \alpha\} \\
&= \sum_{\beta\gamma \in \mathbb{R}} \beta\gamma Pr\{x = \beta, y = \gamma\} \\
&= \sum_{\beta\gamma \in \mathbb{R}} \beta\gamma Pr\{x = \beta\}\,Pr\{y = \gamma\} \\
&= \sum_{\beta \in \mathbb{R}} \beta Pr\{x = \beta\} \sum_{\gamma \in \mathbb{R}} Pr\{y = \gamma\} \\
&= E\,x.E\,y
\end{aligned}
$$

$\square$

Let again x, y be two random variables on the same probability space. Of course we have

$$E(x + y) = E\,x + E\,y. \tag{3.3}$$

If x and y are independent, we have also

$$Var(x + y) = Var\,x + Var\,y \tag{3.4}$$

Indeed, by proposition 3.4

$$
\begin{aligned}
Var(x+y) &= \mathrm{E}\,(x+y)^2 - (\mathrm{E}\,(x+y))^2 \\
&= \mathrm{E}\,x^2 + 2\mathrm{E}\,xy + \mathrm{E}\,y^2 - ((\mathrm{E}\,x)^2 + 2\mathrm{E}\,x\mathrm{E}\,y + (\mathrm{E}\,y)^2) \\
&= \mathrm{E}\,x^2 - (\mathrm{E}\,x)^2 + \mathrm{E}\,y^2 - (\mathrm{E}\,y)^2 \\
&= Var\,x + Var\,y
\end{aligned}
$$

As a straightforward consequence we obtain the following proposition. It expresses the expectation and variance of the random variables of a stochastic process in terms of the expectation or variance of its increments.

Proposition 3.5   . *Let $T \in \mathbb{T}$ and η be a stochastic process indexed by $[0 \dots T]$. Then*

$$
E\eta_T = \sum_{0 \le t < T} E\,\delta\eta_t.
$$

If, in addition, the increments are independent, then

$$
Var\,\eta_T = \sum_{0 \le t < T} Var\,\delta\eta_t.
$$

As already said, if the increments of a stochastic process are given and independent, they implicitly determine the trajectories of the process and their probabilities. It is this possibility which we will use to define our basic stochastic process: the discrete Wiener walk.

3.2   The discrete Wiener walk

Definition 3.6   Let $\delta t > 0$, $T \in \mathbb{T}$. The discrete Wiener walk is the stochastic process W indexed by $[0..T]$ such that W_0 is identically zero, and its increments δW_t are independent and given by

$$
\delta W_t = \begin{cases} \sqrt{\delta t} & \text{with probability } 1/2 \\ -\sqrt{\delta t} & \text{with probability } 1/2 \end{cases}
$$

A trajectory w of the process W is a collection of partial sums; *i.e.*

$$
\begin{aligned}
w : \mathbb{T} &\longrightarrow \mathbb{R} \\
t &\longrightarrow \sum_{s<t} \pm\sqrt{\delta t}.
\end{aligned}
$$

The trajectories are all equiprobable, with probability $\left(\frac{1}{2^{T/\delta t}}\right)$. Indeed, the process W is composed of $2^{T/\delta t}$ trajectories. Note that the process β of the previous section is a Wiener walk for $\delta t = 1$ and $T = 3$.

The Wiener walk is the basic reference for stochastic processes. Under quite general conditions, the trajectories of any stochastic process are close to those of a Wiener walk with close probability distributions. A precise statement of this fact can be found in Chapter 18 of Nelson's book [16]. It was R. M. Anderson in [1] who first got the idea of looking at Wiener walks with infinitesimal steps, and it turned out that this approach enabled to bypass the complications related to the study and understanding of classical, time-continuous stochastic processes such as a thorough knowledge of measure theory, the convergence of stochastic integrals etc ...

The following lemma states that the binomial triangle $\mathcal{C}_{[0..T]}$ is the union of all trajectories of a Wiener walk.

Lemma 3.7 *Let* $(t, x) \in \mathcal{C}$. *Let* $T \in \mathbb{T}$, $T \geq t$. *Then*

(1) *the point* (t, x) *lies on a trajectory of the Wiener walk indexed by* $[0..T]$

(2) *a trajectory of the Wiener walk containing* (t, x) *must contain* $j_{t,x}$ *upward steps and* $\nu_t - j_{t,x}$ *downward steps on* $[0 \ldots t]$ *; hence, the number of upward steps and downward steps up to t of such a trajectory depends only on the values of t and x.*

(3) *there are exactly*

$$
\begin{pmatrix} \nu_t \\ j_{t,x} \end{pmatrix}
$$

trajectories going through the point (t, x).

Proof : Let $\nu = \nu_t$ and $j = j_{t,x}$. Then

$$
x = \left(j - \frac{\nu}{2}\right) \cdot 2\sqrt{\delta t} = j\sqrt{\delta t} + (\nu - j)\left(-\sqrt{\delta t}\right).
$$

So, the point (t, x) is on any trajectory of the Wiener walk which has $j_{t,x}$ upward steps and $\nu_t - j_{t,x}$ downward steps on $[0 \ldots t]$.

The number of those trajectories is equal to the number of subsets of j elements in a set of ν elements, which is $\begin{pmatrix} \nu \\ j \end{pmatrix}$. $\qquad\square$

We will also study stochastic processes which have the same trajectories as the Wiener walk, but with different probabilities. In a precise way, we will consider stochastic processes with the same increments of the Wiener process, but with probability p -instead of $1/2$- for an upward step and probability $1 - p$ for a downward step, with $0 < p < 1$. Generally, the number p will be of the form $p = \frac{1}{2} + a\sqrt{\delta t}$, for some $a \in \mathbb{R}$, and therefore we will write $p = p_a$.

Definition 3.8 Let $T \in \mathbb{T}$ and $a \in \mathbb{R}$ such that $0 < p_a < 1$. We will denote by $W^{(a)}$ the process indexed by $[0..T]$, with $W_0^{(a)} = 0$ and with independent increments satisfying

$$\delta W_t^{(a)} = \begin{cases} \sqrt{\delta t} & \text{with probability } p_a \\ -\sqrt{\delta t} & \text{with probability } 1 - p_a \end{cases}$$

We determine now the probability distributions of these processes.

Theorem 3.9 Let $T \in \mathbb{T}$ and $a \in \mathbb{R}$ such that $0 < p_a < 1$. For all $t \leq T$, the probability that a trajectory of the process $W^{(a)}$ contains the point (t, x) of the binomial cone $\mathcal{C}$ is

$$\Pr\left\{ W_t^{(a)} = x \right\} = b_a(t, x)\delta x.$$

Proof : Let $t \in [0..T]$. Put $p = p_a$, $\nu = \nu_t$ and $j = j_{t,x}$. The number of classes $\overline{\lambda}_t$ (see Notation 3.2 and Figure 3.1) is equal to the number of trajectories of the Wiener walk indexed by $[0..t]$ which end at the point x, and by Lemma 3.7, this number is equal to $\begin{pmatrix} \nu_t \\ j_{t,x} \end{pmatrix}$. Using the independence, the probability of each class is then

$$\Pr\overline{\lambda}_t = \prod_{0 \leq s < t} \Pr\left\{ \delta W_s^{(a)} = \delta\lambda(s) \right\} = p^j(1 - p)^{\nu - j}$$

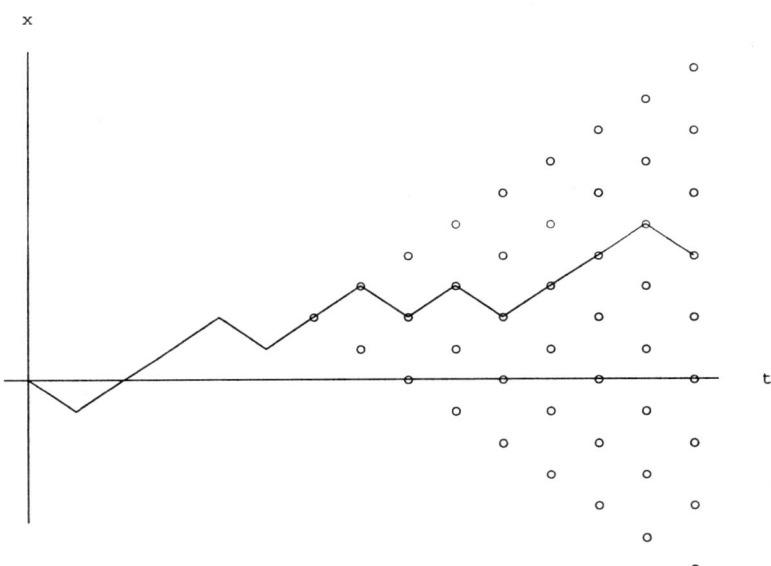

Fig. 3.1   A class $\bar{\lambda}_t$ in a Wiener walk with $t = 6\delta t$

Hence,

$$\Pr\left\{W_t^{(a)} = x\right\} = \sum_{\bar{\lambda}} \Pr\left(\bar{\lambda}\right) = \left(\begin{array}{c} \nu_t \\ j_{t,x} \end{array}\right) p^j (1-p)^{\nu-j} = b_a(t,x)\delta x.$$

$\square$

Corollary 3.10   *Let $T \in \mathbb{T}$, $(t,x) \in \mathcal{C}$, $t \leq T$ and W be the Wiener walk indexed by $[0..T]$. Then $\Pr\{W_t = x\} = b(t,x)\delta x$.*

The next theorem states that if the parameter a is limited, the probability distributions of the processes $W^{(a)}$, including the Wiener walk $W^{(0)}$, are nearly normal.

Theorem 3.11   *Let $\delta t \simeq 0$, let a be limited and let $T \in \mathbb{T}$, $t \in [0..T]$ be appreciable. Let $W^{(a)}$ be indexed by $[0..T]$. Then*

$$\Pr\left\{W_t^{(a)} \leq x\right\} \simeq \mathcal{N}_{2at,\sqrt{t}}(x).$$

Proof : Suppose that x is limited. Applying theorem 2.19 to the indicator function of the interval $]-\infty, x]$ we get

$$
\begin{aligned}
\Pr\left\{W_t^{(a)} \leq x\right\} &= \sum_{-t/\delta t \leq y \leq x} b_a(t,y)\delta x \\
&\simeq \frac{1}{\sqrt{2\pi t}} \int_{-\infty}^{x} \exp\left(-\frac{(y-2at)^2}{2t}\right) dy \\
&= \mathcal{N}_{2at,\sqrt{t}}(x).
\end{aligned}
$$

If x is unlimited, the near equality remains valid because the masses of both probability distributions contain only limited points: if x is positive then both the probability distributions are nearly 0, and if x is negative then both are nearly equal to 1.   □

If the time step δt is infinitesimal, on a miscroscopic scale the Wiener walk has discrete properties: concrete, infinitesimal increments. On a macroscopic scale it has continuous properties. Indeed, by the above theorem its probability distribution is nearly normal, which implies that it is S-continous, the binomial function $b_a(t,x)$, also S-continuous, acting as a "discrete density". The trajectories are discrete, finite sequences, but on a macroscopic level they mostly are S-continuous. A precise statement and proof of this important property is given in Chapter 13 of [16]. An example of a S-continuous trajectory is given by a trajectory wich has alternately positive and negative increments; see exercise 3.3. This exercise also provides an exemple of a trajectory which is really discrete, i.e. not S-continuous. The above remarks lead to the following fundamental observation: there are properties of the Wiener walk which do *not* depend on the the actual lenght of δt, as long as it is infinitesimal. Such properties are sometimes called *macroscopic*; they are usually external, using expressions like "is nearly equal to", "is S-continuous", "holds except on a set of infinitesimal probability". It is just these properties which have counterparts within the theory of continuous-time stochastic processes, but we will not develop this point.

Here are two "macroscopic properties" of the Wiener walk, which are a consequence of Theorem 3.11 and the properties of the Normal Distribution: the expectation of the random variable $W_t^{(a)}$ is equal to $2at$ and its variance is approximately equal to t. On a microscopic level the variances depend on δt. This is seen, if we calculate these quantities exactly.

Proposition 3.12 *Let $t, T \in \mathbb{T}$ with $t \le T$ and $W^{(a)}$ be indexed by $[0\ldots T]$. Then*

$$
\begin{aligned}
EW_t^{(a)} &= 2at & (3.5) \\
VarW_t^{(a)} &= t(1 - 4a^2\delta t) & (3.6)
\end{aligned}
$$

Proof : The formulae are trivially true for $t = 0$. Let $t > 0$. Notice that

$$
\mathrm{E}\,\delta W_t^{(a)} = \sqrt{\delta t}(1/2 + a\sqrt{\delta t}) + (-\sqrt{\delta t})(1/2 - a\sqrt{\delta t}) = 2a\delta t.
$$

Then by proposition 3.5

$$
\begin{aligned}
\mathrm{E}\,W_t^{(a)} &= \sum_{0 \le s < t} \mathrm{E}\,\delta W_s^{(a)} \\
&= \sum_{0 \le s < t} 2a\delta t \\
&= 2at
\end{aligned}
$$

and

$$
\begin{aligned}
VarW_t(a) &= \sum_{0 \le s < t} Var\delta W_s^{(a)} \\
&= \sum_{0 \le s < t} \mathrm{E}\,(\delta W_s^{(a)})^2 - (\mathrm{E}\,\delta W_s^{(a)})^2 \\
&= \sum_{0 \le s < t} \delta t(1/2 + a\sqrt{\delta t}) + \delta t(1/2 - a\sqrt{\delta t}) - 4a^2(\delta t)^2 \\
&= \sum_{0 \le s < t} \delta t(1 - 4a^2\delta t) \\
&= t(1 - 4a^2\delta t)
\end{aligned}
$$

$\square$

N.B. An alternative proof may be based on the combinatorical exercise 1.3

In the following sections we will use the Wiener walk to illustrate some fundamental notions related to stochastic processes.

3.3   Recombining processes; the discrete surface

The Wiener walk presents the following two characteristics:

(1) at any time, the next step is the result of one of two possible movements: it is an upward step or a downward step
(2) an upward step followed by a downward step leads to the same position as a downward step followed by an upward step.

A stochastic process the trajectories of which present the first characteristic is called *bivalent*. If they possess also the second characteristic of *commutativity* the process is called *recombining*. This is formalized in the following definition.

Definition 3.13   Let $T \in \mathbb{T}$ and η be a finite stochastic process indexed by $[0..T]$. We assume that η_0 is constant.

(1) The process η is said to be *bivalent* if for any trajectory λ of η and for any $t < T$, the set $\overline{\lambda}_t$ is the union of two non-empty, disjoint sets of trajectories: a set $\overline{\lambda}_t^+$ of trajectories which, at time $t + \delta t$, make a step u, and a set $\overline{\lambda}_t^-$ of trajectories which, at time $t + \delta t$, make a step d, where $d < u$.
(2) An element $\alpha \in \overline{\lambda}_t^+$ will be called *an upper prolongation of* λ at time t, and $\delta\alpha_t = u$ will be called *the upper movement of* α at time t. An element $\beta \in \overline{\lambda}_t^-$ will be called *a lower prolongation of* λ at time t, and $\delta\beta_t = d$ will be called *the lower movement of* α at time t.
(3) A process η is called *recombining* if

- η is bivalent
- at any time, two trajectories which have the same number of upward movements and downward movements assume the same value.

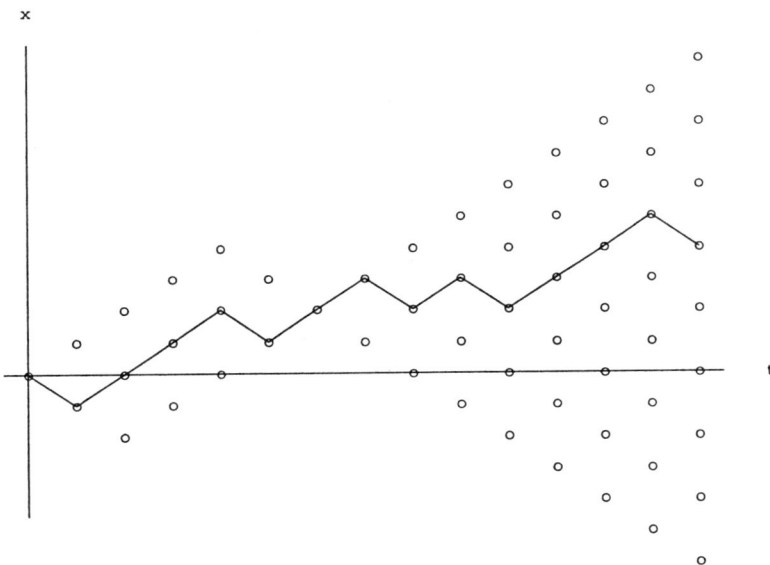

Fig. 3.2 The class $\Lambda_{t,x}$ for $t = 6\delta t$ and $x = 2(\delta t)^{\frac{1}{2}}$

The configuration of a bivalent process is tree-like.

The configuration of the trajectories of a recombining process is a collection of diamonds. To see this, we use the following observation on the prolongation of trajectories.

Let α be a trajectory of η. Assume that at time $t \leq T - 2\delta t$ it has an upper movement and at time $t + \delta t$ it has a lower movement. Let β be a second trajectory. Assume that the number of upward movements of β up to t is equal to the number of upward movements of α up to t. Assume further that β has a downward movement at T and an upward movement at time $t + \delta t$. Then $\alpha(t + \delta t) > \beta(t + \delta t)$ and the number of upward movements of both trajectories up to time $t + 2\delta t$ being equal, we have $\alpha(t + 2\delta t) = \beta(t + 2\delta t)$. So indeed, after their separation at $t + \delta t$ the two trajectories recombine.

In the case of the random Wiener walk, the diamonds are identical, but this is not so for all recombining processes. In Chapter 3.8 we will introduce the geometric Brownian motion; it is a recombining process were the size and form of the diamonds depends on time and space.

Let η be a recombining process indexed by $[0..T]$ for some $T \in \mathbb{T}$. In the following, we will explain how the process η may be seen as a real-valued mapping defined on the binomial triangle $\mathcal{C}_{[0..T]}$. This will allow us to see the collection of all the values taken by the trajectories of the process η as a discrete surface of $\mathbb{R}^3$ defined on a binomial triangle $\mathcal{C}_{[0..T]}$.

We first introduce some notations.

Notation : Let $T \in \mathbb{T}$ and $(t, x) \in \mathcal{C}_{[0..T]}$. Let η be a recombining process indexed by $[0..T]$. We denote by $\Lambda_{t,x}$ the collection of all trajectories which at time t contain $j_{t,x}$ upward steps.

Note that if a trajectory at time t has $j_{t,x}$ upward steps, it has $\nu_t - j_{t,x}$ downward steps. Hence $\Lambda_{t,x}$ is the set of all trajectories which up to time t have $j_{t,x}$ upward steps and $\nu_t - j_{t,x}$ downward steps. Since the process is recombining, all elements of $\Lambda_{t,x}$ take the same value at time t; in the case of the Wiener walk, we have $\overline{\lambda}_t = x$ for any $\lambda \in \Lambda_{t,x}$. This observation justifies the following definition.

Definition 3.14 Let $T \in \mathbb{T}$, and let η be a recombining process indexed by $[0..T]$. The *discrete surface associated to* η is the graph of the mapping $\tilde{\eta} : \mathcal{C}_{[0..T]} \longrightarrow \mathbb{R}$ defined by $\tilde{\eta}(t, x) = \lambda(t)$ for any $\lambda \in \Lambda_{t,x}$.

Examples : Consider the random walk u defined by

$$
\left\{
\begin{array}{ll}
u(0) & = \quad 0 \\
\\
\delta u_t & = \quad \left\{
\begin{array}{ll}
\sigma\sqrt{\delta t} & \text{with probability } \frac{1}{2} \\
\\
-\sigma\sqrt{\delta t} & \text{with probability } \frac{1}{2}
\end{array}
\right.
\end{array}
\right.
$$

for $\sigma > 0$. Then, for each $(t, x) \in \mathcal{C}$ we have

$$
\begin{aligned}
\tilde{u}(t, x) & = j_{t,x}\sigma\sqrt{\delta t} + (\nu_t - j_{t,x})(-\sigma\sqrt{\delta t}) \\
& = (2j_{t,x} - \nu_t)\sigma\sqrt{\delta t} \\
& = \sigma x
\end{aligned}
$$

The discrete surface associated to the process v defined by

$$\begin{cases} v(0) &= 0 \\ \delta v_t &= t\delta t + \delta u_t \end{cases}$$

satisfies the equation

$$\tilde{v}(t,x) = \sum_{0 \le s < t} s\delta t + \sigma x.$$

If $\delta t \simeq 0$ and t is limited, we get the approximation

$$\tilde{v}(t,x) \simeq \frac{t^2}{2} + \sigma x.$$

The discrete surface associated to the geometric Brownian motion which we will study in Chapter 3.8 is an exponentially curved surface in both directions t and x.

A discrete surface is in one-to-one correspondence with the binomial triangle. Indeed, taking time-sections, we see that they have the same geometry: they consist of one point at time zero, two points at time δt, three points at time $2\delta t$, up to $T/\delta t + 1$ points at time T.

In the remaining part of this study, we will frequently identify the stochastic process η with its associated mapping $\tilde{\eta}$ and, somewhat abusively, we will see the process η not as being indexed by the interval $[0..T] \subset \mathbb{T}$, but as being defined on the binomial triangle $\mathcal{C}_{[0..T]}$.

Within such identifications, a random variable η_t becomes a section $(t, \eta(t, \cdot))$ of the discrete surface. Also, a trajectory λ of the process becomes a discrete curve $t \mapsto \eta(t, x(\lambda, t))$ on the surface. When it reaches the point $\eta(t, x)$, we say that its space coordinate $x(\lambda, t)$ within the binomial triangle is the *state* of the trajectory. If there is no ambiguity with respect to the time we write $x(\lambda)$ instead of $x(\lambda, t)$. So $\Lambda_{t,x}$ is the set of all trajectories which at time t are in the state x: all trajectories which at time t are in the state x have the same number $j_{t,x}$ of upward steps and the same number $\nu_t - j_{t,x}$ downward steps.

Consider a trajectory which at time t is in the state x. On the discrete surface, the two possible prolongations of the curve corresponding to this trajectory are $\eta\left(t + \delta t, x + \sqrt{\delta t}\right)$ and $\eta\left(t + \delta t, x - \sqrt{\delta t}\right)$; these prolongations may combine in $\eta(t + 2\delta t, x)$.

We end this section by introducing some notations.

Notations :

Let $T \in \mathbb{T}$, and let η be a recombining process defined on $\mathcal{C}_{[0..T]}$. Let $(t, x) \in \mathcal{C}_{[0..T]}$ be such that $t < T$. We write

- $\eta(t, x)^+ \;=\; \eta\left(t + \delta t, x + \sqrt{\delta t}\right)$, $\bullet$ $\delta\eta(t, x)^+ = \eta(t, x)^+ - \eta(t, x)$
- $\eta(t, x)^- \;=\; \eta\left(t + \delta t, x - \sqrt{\delta t}\right)$, $\bullet$ $\delta\eta(t, x)^- = \eta(t, x)^- - \eta(t, x)$
  - $\Lambda_{t,x}^+ \;=\; \left\{\alpha \in \Lambda_{t,x} \;\bigm|\; \delta\alpha_t = \delta\alpha(t, x)^+\right\}$
  - $\Lambda_{t,x}^- \;=\; \left\{\alpha \in \Lambda_{t,x} \;\bigm|\; \delta\alpha_t = \delta\alpha(t, x)^-\right\}$
  - $p_{t,x} \;=\; \Pr \Lambda_{t,x} / \Pr \Lambda_{t,x}^+$

Let λ be a trajectory of η. Notice that we have

- $\overline{\lambda}_t^+ \;=\; \left\{\alpha \in \overline{\lambda}_t \;\bigm|\; \delta\alpha_t = \delta\eta(t, \lambda(t))^+\right\}$
- $\overline{\lambda}_t^- \;=\; \left\{\alpha \in \overline{\lambda}_t \;\bigm|\; \delta\alpha_t = \delta\eta(t, \lambda(t))^-\right\}$

3.4 Conditional probability and conditional expectation

We start this section be recalling the classical notions of conditional probability and conditional expectation in a finite probability space. After that, we will extend these notions to stochastic processes, in which case, conditional probabilities and conditional expectations will themselves become stochastic processes. Often we restrict our study to recombining processes taking advance of their geometric properties. More general definitions of conditional probability and conditional expectations can be found in Nelson's book for example.

Let (Ω, pr) be a finite probability space and let $A \subset \Omega$. Let $\omega \in A$. The *conditional probability* $\mathrm{pr}_A(\omega)$ *of* ω *knowing* A is defined by

$$\mathrm{pr}_A(\omega) = \frac{\Pr(\{\omega\} \cap A)}{\Pr A}. \qquad (3.7)$$

Let $x : \Omega \longrightarrow \mathbb{R}$ be a random variable. The *conditional expectation* $E_A x$

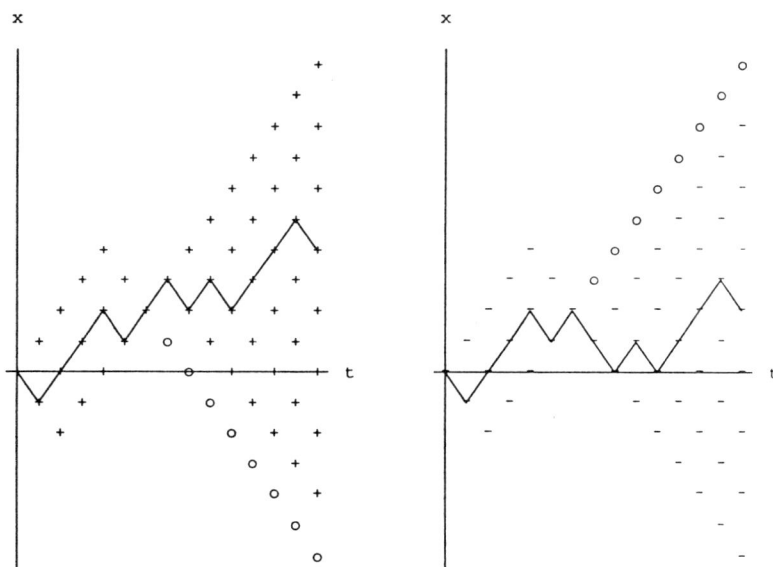

Fig. 3.3 Projection on the binomial network of the class $\Lambda_{t,x}^+$, which is indicated by the + signs (on the left), and that of the class $\Lambda_{t,x}^-$, which is indicated by the − signs (on the right); the small circles represent the points which must be added to the network if we want to get the whole of the class $\Lambda_{t,x}$. In this figure, $t = 6\delta t$ and $x = 2(\delta t)^{\frac{1}{2}}$. To each trajectory $\alpha \in \Lambda_{t,x}^+$, there corresponds a trajectory $\beta = \beta(\alpha) \in \Lambda_{t,x}^-$ with exactly the same movements, except for the movement at time $t = 4\delta t$, which is an upward movement for α and a downward movement for β.

knowing A is defined by

$$\mathrm{E}_A x = \sum_{\alpha \in A} x(\alpha) \mathrm{pr}_A(\alpha). \qquad (3.8)$$

One also uses the notation $\mathrm{pr}\,(\omega|A)$ and $\mathrm{E}\,(x|A)$.

As an example, let Ω be the set of students of some course, and $\mathcal{A}$ be the set of classes. Let $A_n \in \mathcal{A}$ be the nth class. Suppose $x : \Omega \to \mathbb{R}$ represents the result of student ω at the exam, which is held simultanously for all students. Giving each individual the same weight $\mathrm{E}\,x$ represents the mean result for the whole cohort, while $\mathrm{E}_A x$ represents the mean of the students belonging to the same class A. Continuing our example we will see that the conditional expectation may act as an random variable. Suppose the supervisor S is confronted with the poor performance of student ω. Then it is of some interest to know whether the performance is poor as an individual indeed, or whether it is possibly influenced by ω^s peers. Then S proceeds as follows:

(1) The class $A_{n(\omega)}$, which contains ω is determined.
(2) On the set $A_{n(\omega)}$ the conditional expectation $A_{n(\omega)}x$ is calculated with the aid of formula (3.8).

This operation is summerized by the definition of the random variable

$$\mathrm{E}_A x(\omega) \equiv \mathrm{E}_{A_n(\omega)} x.$$

The conditional expectation may be defined in this manner on every family of subsets forming a partition of a finite probability space. We will consider two kinds of partitions: classes of trajectories with a common history and classes of trajectories with a common state.

Let Ω be the set Λ of trajectories of a stochastic process. The next definition expresses the probability of a trajectory, knowing that we have followed it up to some time, and in the same circumstances, the expectation of a random variable defined on the set of trajectories.

Definition 3.15 Let $T \in \mathcal{T}$ and η be a stochastic process indexed by $[0..T]$. Let $g : \Lambda \longrightarrow \mathbb{R}$ be a random variable. We call the *trajectorial conditional probability* p_t and the *trajectorial conditional expectation* $E_t g$ at time t the random variables

$$
\begin{aligned}
p_t(\lambda) &= pr_{\overline{\lambda}_t}(\lambda) \\
E_t g(\lambda) &= E_{\overline{\lambda}_t} g
\end{aligned}
$$

Notice that we have

$$
\begin{aligned}
p_t(\lambda) &= pr(\lambda)/Pr(\overline{\lambda}_t) \\
E_t g(\lambda) &= \sum_{\alpha \in \overline{\lambda}_t} g(\alpha) pr_t(\alpha) = \frac{1}{Pr(\overline{\lambda}_t)} \sum_{\alpha \in \overline{\lambda}_t} g(\alpha) pr(\alpha)
\end{aligned}
$$

Also, if $\beta \in \overline{\lambda}_t$ is arbitrary, we have $E_t g(\beta) = E_t g(\lambda)$: the mean being taken over the set of all trajectories with a common history must of course be the same for any of its members. For a better understanding of the above notions, we present some examples.

Examples : Let β be the process which we considered in the example of Section 3.1 with $\delta t = 1$.

- For any trajectory λ we have $pr(\lambda) = 1/8$, $pr_1(\lambda) = 1/4$, $pr_2(\lambda) = 1/2$ and $pr_3(\lambda) = 1$.
- Let λ be a trajectory of β. Then, using the conditional probabilities above,

$$
\begin{aligned}
E_3 \beta_3(\lambda) &= \lambda(3) \\
E_2 \beta_3(\lambda) &= \frac{1}{2}(\lambda(2) + 1) + \frac{1}{2}(\lambda(2) - 1) = \lambda(2) \\
E_1 \beta_3(\lambda) &= \frac{1}{4}(\lambda(1) + 2) + \frac{2}{4}\lambda(1) + \frac{1}{4}(\lambda(1) - 2) = \lambda(1) \\
E_0 \beta_3(\lambda) &= E\beta_3 = 0
\end{aligned}
$$

We see that $E_t \beta_3(\lambda) = \lambda(t)$; generally speaking as time increases the conditional expectation of a random variable becomes closer to its true value at time T.

The following useful proposition expresses the conditional probability of a trajectory in terms of the conditional probabilities of its future increments.

Proposition 3.16 *Let $T \in \mathbb{T}$. Let η be a stochastic process indexed by $[0 \ldots T]$. Then for every trajectory λ and every t with $0 \le t < T$*

$$
pr_t(\lambda) = \prod_{t \le s < T} Pr_s \{\delta\eta_s = \delta\lambda(s)\} \tag{3.9}
$$

Proof : Notice that for all $s > 0$

$$
\overline{\lambda}_t = \{(\forall \tau \le s - \delta t)\, \delta\eta_\tau = \delta\lambda(\tau)\}
$$

It follows from the definition of the conditional probability that as long as $t > 0$ and $s \in [t \ldots T - \delta t]$

$$Pr_s \{\delta\eta_s = \delta\lambda(s)\} = \frac{Pr\{(\forall \tau \leq s)\, \delta\eta_\tau = \delta\lambda(\tau)\}}{Pr\{(\forall \tau \leq s - \delta t)\, \delta\eta_\tau = \delta\lambda(\tau)\}}$$

So, using the calculation principle of formula (2.1) we obtain

$$\prod_{t \leq s < T} Pr_s \{\delta\eta_s = \delta\lambda(s)\} \;=\; \frac{Pr\{(\forall \tau \leq T)\, \delta\eta_\tau = \delta\lambda(\tau)\}}{Pr\{(\forall \tau < t)\delta\eta_\tau = \delta\lambda(\tau)\}}$$

$$= \frac{pr\lambda}{Pr\overline{\lambda}_t}$$

$$= pr_t(\lambda)$$

Because trivially

$$Pr_0 \{\delta\eta_0 = \delta\lambda(0)\} = Pr\{(\forall \tau < \delta t)\delta\eta_\tau = \delta\lambda(\tau)\}$$

and $pr_0\lambda = pr\lambda$ it follows from the above that formula (3.9) also holds for $t = 0$. $\square$

Let A be an event and g be a random variable. Notice that if we let vary t the sequence of conditional probabilities $(Pr_t A)_{t \in [0 \ldots T]}$ and the sequence of conditional expectations $(E_t g)_{t \in [0 \ldots T]}$ become a stochastic process.

The above definition concerned probabilities and expectations at time t conditioned to the knowledge of the whole past of a trajectory up to time t.

It is also interesting to consider probabilities and expectations conditioned to the mere knowledge of the position at time t : only the actual situation matters, not the history. For convenience we formulate these notions for recombining processes.

Definition 3.17   Let $T \in \mathbb{T}$ and η be a recombining process defined on $\mathcal{C}_{[0 \ldots T]}$. Let us denote by Ω the set of all trajectories of η, and let us consider the event $\Lambda_{t,x} \subset \Omega$. We will write $pr_{t,x}$ instead of $pr_{\Lambda_{t,x}}$, so that if $(t, x) \in \mathcal{C}_{[0 \ldots T]}$ and if $\lambda \in \Lambda_{t,x}$ then

$$\operatorname{pr}_{t,x}(\lambda) = \frac{\operatorname{pr}(\lambda)}{\operatorname{Pr}(\Lambda_{t,x})}. \tag{3.10}$$

- We will call $\operatorname{pr}_{t,x}$ the *conditional probability at time t and state x.*

- Let $f : \Omega \longrightarrow \mathbb{R}$ be a random variable. The *conditional expectation of f at time t and state x* is by definition

$$\operatorname{E}_{t,x} f = \sum_{\lambda \in \Lambda_{t,x}} f(\lambda) \operatorname{pr}_{t,x}(\lambda). \tag{3.11}$$

- The *conditional variance of f at time t and state x* is by definition

$$\operatorname{Var}_{t,x} f = \operatorname{E}_{t,x}(f - \operatorname{E}_{t,x} f)^2$$

- The *conditional standard deviation of f at time t and state x* is by definition the square-root of its state conditional variance.

Examples : Let β be again the process which we considered in the example of Section 3.1 with $\delta t = 1$.

- Usually the ordinary probability and the conditional probability are different. We have for example,

$$\operatorname{Pr}\{\beta_3 = 1\} = \frac{3}{8},$$

while

$$\operatorname{Pr}_{2,0}\{\beta_3 = 1\} = \frac{\operatorname{Pr}\{\beta_3 = 1, \beta_2 = 0\}}{\operatorname{Pr}\Lambda_{2,0}} = \frac{\frac{1}{4}}{\frac{1}{2}} = \frac{1}{2}.$$

If λ is a trajectory going through $(2,0)$ then

$$\operatorname{pr}\lambda = \frac{1}{8}$$

while

$$\operatorname{pr}_{2,0}\lambda = \frac{\operatorname{pr}\lambda}{\operatorname{Pr}\Lambda_{2,0}} = \frac{1/8}{1/2} = \frac{1}{4}.$$

- The ordinary probabilities and the conditional probabilities of the increments of the process β, and in general the discrete Wiener walk coincide. For example, at any time t we have pr $\{\delta\beta_t = 1\} = \frac{1}{2}$; but also at any time pr$_{t,x}$ $\{\delta\beta_t = 1\} = \frac{1}{2}$; if, for example, $t = 2$ and $x = 0$ then

$$\text{pr}_{2,0}\{\delta\beta_2 = 1\} = \frac{\text{Pr}\{\delta\beta_2 = 1, \beta_2 = 0\}}{\text{Pr}\,\Lambda_{2,0}} = \frac{\frac{1}{4}}{\frac{1}{2}} = \frac{1}{2}$$

See also formula (3.12) below.

- The conditional expectation of β_3 at time $t = 2$ and for $x = 0$ is

$$\text{E}_{2,0}\beta_3 = \sum_{\lambda \in \Lambda_{2,0}} \beta_3(\lambda)\text{pr}_{2,0}\lambda = 1\cdot\frac{1}{4} + 1\cdot\frac{1}{4} - 1\cdot\frac{1}{4} - 1\cdot\frac{1}{4} = 0$$

Also,

$$\begin{aligned}
\text{E}_{2,2}\beta_3 &= \sum_{\{\lambda \mid \lambda(2)=2\}} \beta_3(\lambda)\text{pr}_{2,2}(\lambda) = \sum_{\{\lambda \mid \lambda(2)=2\}} \beta_3(\lambda)\frac{\text{pr}\,(\lambda)}{\text{pr}\,\Lambda_{2,2}} \\
&= 3\cdot\frac{1/8}{1/4} + 1\cdot\frac{1/8}{1/4} = 2
\end{aligned}$$

As a generalization one may obtain for all t and x

$$\text{E}_{t,x}\beta_3 = x$$

So we have $E_{t,.}\beta_3 = E_t\beta_3$ for all t and x. Theorem 3.33 says that the trajectorial conditional expectation and the state conditional expectation coincide for a certain general class of random variables; that they do not always coincide is shown by exercise 11.

As it was the case of the conditional expectation E_t, the conditional expectation $E_{t,x}$ gives a more precise estimation of the random variable than the non-conditional expectation: we make use of additional information on the state of the process at time t. For example, we have $E_{2,2}\beta_3 = 2$, a value which is nearer to the possible values of the random variable β_3 in case at time $t = 2$ the processs reaches the value 2, than $E\,\beta_3 = 0$ – which is the expectation on all the values that may be taken by β_3– .

The functions defined in definition 3.17 may be interpreted as discrete surfaces. As explained in the previous paragraph their sections corresponding to a fixed time t may then be interpreted as random variables. For instance, the state conditional expectation $E_{t,.}$ of a given random variable g could explicitly be defined as a random variable by

$$E_{t,.}g(\lambda) = E_{t,x(\lambda)}g$$

If we let vary t, the sequence of state conditional expectations becomes a (recombining) stochastic process.

So we met two different situations, where the conditional expectation is to be considered as a stochastic process: the family of trajectorial conditional expectations and the family of state conditional expectations. In fact the conditional expectation becomes a stochastic process as a result of more general constructions. We do not go further is this direction, and refer instead to Nelson [16, chapter 9].

3.5 Markov processes, binomial processes

An important observation on stock-price processes is that normally all information on the issuing company is discounted in the actual stock-price. As is indicated in Chapter 4, this observation is incorporated into the assumptions of the usual models of the developments in time of stock-prices. The assumption implies that knowledge of the actual state of the process is equivalent to the knowledge of its past: i.e. knowledge of the past does not lead to more precise predictions. This property is known as the *Markov* property, and processes which possess this property are known as *Markov* processes. Below we introduce the notion of Markov process in the geometrical context of recombining processes; we remark that the notions of de the definition below are special versions of notions which are usually given in the context of more general stochastic processes (see for example [11]).

Definition 3.18

Let $T \in \mathbb{T}$, and let η be a recombining process defined on $\mathcal{C}_{[0..T]}$. The process η will be called a *Markov process* if for any $(t, x) \in \mathcal{C}_{[0..T]}$ such that $t < T$ and for any $\lambda \in \Lambda_{t,x}$

$$\Pr{}_t\left\{\delta\eta = \delta\eta(t,x)^+\right\} = \Pr{}_{t,x}\left\{\delta\eta_t = \delta\eta(t,x)^+\right\}(\equiv p_{t,x})$$

Notice that the downward movements of a Markov process automatically satisfy

$$\Pr{}_t\left\{\delta\eta = \delta\eta(t,x)^-\right\} = \Pr{}_{t,x}\left\{\delta\eta_t = \delta\eta(t,x)^-\right\}(= 1 - p_{t,x})$$

Examples :

- Let $T \in \mathbb{T}$. Consider the Wiener walk indexed by $[0..T]$. For any $(t,x) \in \mathcal{C}$ such that $t < T$, we have

$$\Pr{}_{t,x}\left\{\delta W_t = \sqrt{\delta t}\right\} \;=\; 1/2 \tag{3.12}$$

$$\Pr{}_{t,x}\left\{\delta W_t = -\sqrt{\delta t}\right\} \;=\; 1/2. \tag{3.13}$$

Indeed, all trajectories are equiprobable, and the number of elements of $\Lambda_{t,x}^+$ is equal to the number of elements of $\Lambda_{t,x}^-$. Hence $Pr\Lambda_{t,x}^+ = Pr\Lambda_{t,x}^-$ and $Pr\Lambda_{t,x}^+/Pr\Lambda_{t,x} = 1/2$. This implies that

$$\Pr{}_{t,x}\left\{\delta W_t = \sqrt{\delta t}\right\} = \Pr{}_{t,x}\left\{\delta W_t = -\sqrt{\delta t}\right\} = 1/2.$$

We show now that the Wiener walk is a Markov process. Let $\lambda \in \Lambda_{t,x}$. We know that all trajectories are equiprobable and that there are as many elements in $\overline{\lambda}_t^+$ than in $\overline{\lambda}_t^-$. Hence $Pr\,\overline{\lambda}_t^+ = Pr\,\overline{\lambda}_t^-$ and $Pr\,\overline{\lambda}_t^+/Pr\,\overline{\lambda}_t = 1/2$. This implies that, also

$$Pr_t\left\{\delta W_t = \sqrt{\delta t}\right\} = Pr_t\left\{\delta W_t = -\sqrt{\delta t}\right\} = 1/2$$

- The geometric Brownian motion of Chapter 3.8 is a Markov process: to see that, one may apply the same reasoning as above or use Proposition 3.24 below.
- In Exercise 12, we will give an example of a recombining process which is not a Markov process.

The notion of Markov-process is obviously related to the notion of independence. Indeed, if the increments of a process are independent, the past of a trajectory -*i.e.* the family of its successive increments in time- tells nothing about the possible development of the process in the immediate future. We see that the Markov-property is somewhat weaker: the past tells nothing more than the actual state. This corresponds to a weaker notion of independence, called *conditional independence*:

Definition 3.19 Let $T \in \mathbb{T}$, and let η be a recombining process defined on $\mathcal{C}_{[0..T]}$.

We will say that the increments of η are *conditionally independent* if for any trajectory λ we have

$$\mathrm{pr}\,\lambda = \prod_{0 \leq t < T} \mathrm{Pr}_{t,\lambda(t)} \{\delta\eta_t = \delta\lambda(t)\}.$$

Clearly the increments of the Wiener walk are conditionally independent. The increments of a stochastic process tend to be dependent if their size depends on the actual value of the trajectories, but then they may still be conditionally independent. This will be the case of the geometric Brownian motion of Chapter 4 (see also Exercise 9).

Proposition 3.20 *Let $T \in \mathbb{T}$. Let η be a recombining Markov process defined on $\mathcal{C}_{[0..T]}$. Then the increments of η are conditionally independent.*

Proof : We have, using proposition 3.16

$$\begin{aligned}
\mathrm{pr}\,\lambda &= \prod_{0 \leq t < T - \delta t} \mathrm{Pr}_t \{\delta\eta_t = \delta\lambda(t)\} \\
&= \mathrm{Pr}_{t,\lambda(t)} \{\delta\eta_t = \delta\lambda(t)\}
\end{aligned}$$

$\square$

We now have a new way to define a recombining stochastic process implicitly:

(1) We give its increments and the corresponding conditional probabilities.

(2) We require either that it is a process with conditionally independent increments, or that it is a Markov process.

In this way, binomial processes are defined as follows:

Definition 3.21 Let $T \in \mathbb{T}$ and $0 < p < 1$. Let η be a recombining process defined on $\mathcal{C}_{[0..T]}$. Then the process η is said to be a *binomial process with parameter p* if

(1) its increments are conditionally independent
(2) for any $(t, x) \in \mathcal{C}_{[0..T]}$ such that $t < T$ we have $p_{t,x} = p$

It results from this definition that if η is a binomial process then

$$\text{pr}_{t,x}\{\delta\eta_t = \delta\eta(t, x)^-\} = 1 - p.$$

The above procedure is also applied to recognize certain mathematical structures as processes. This will be done notably in Chapter 5. We will endow the hedging procedure of an option on a stock with a conditional probability, thus interprete it as a stochastic process – in fact a binomial process – , after which the option-price will have the form of an expectation, a binomial sum to be calculated with the aid of theorem 2.19.

Next proposition gives the probabilities of the trajectories λ and the classes $\overline{\lambda}_t$ and $\Lambda_{t,x}$. Its proof is similar to that of Theorem 3.9 and is elementary from the combinatorial point of view.

Proposition 3.22   *Let $T \in \mathbb{T}$ and $0 < p < 1$, and let η be a binomial process defined on $\mathcal{C}_{[0..T]}$ with parameter p.*

- *If λ is a trajectory of η then*

$$pr\,\lambda = p^j(1 - p)^{\nu-j},$$

  where $\nu \equiv \nu_T$ is the number of increments of the process and $j \equiv j_{T,\lambda(T)}$ is the number of upward steps of the trajectory λ.
- *If λ is a trajectory of η and $t \in T$ then*

$$\Pr(\overline{\lambda}_t) = p^j(1 - p)^{\nu_t-j},$$

  where ν_t is the number of increments of the process up to t and $j \equiv j_{t,\lambda(t)}$ is the number of upward steps of the trajectory λ up to t.
- *Let $(t, x) \in \mathcal{C}_{[0..T]}$ with $t < T$, and let $a \in \mathbb{R}$ be such that $p = p_a$. Then*

$$\Pr \Lambda_{t,x} = b_a(t, x)\delta x.$$

If a and $\lambda(T)$ are limited we may compare the probabilities of a trajectory λ of a binomial process with parameter $p = p_a$ and parameter $p = 1/2$.

This is done in the next proposition which is a very elementary form of a general theorem on change of probabilities, called *Girsanov's theorem*.

Proposition 3.23   *Let $T \in \mathbb{T}$ and η be a binomial process with parameter p_a. Let $\lambda \in \Lambda(T, x)$. Assume a and T are limited. Then for all $x \in \mathcal{C}$*

$$pr(\lambda) = (1/2)^T \cdot exp(-2a^2 T + (2a + \oslash)x + \oslash) \qquad (3.14)$$

Proof :   We have using proposition 3.22 and Euler's formula

$$
\begin{aligned}
pr(\lambda) &= p_a^{j_{t,x}}(1 - p_a)^{\nu_T - j_{t,x}} \\
&= (1/2 + a\sqrt{\delta t})^{T/(2\delta t) + x/\delta x}(1/2 - a\sqrt{\delta t})^{T/(2\delta t) - x/\delta x} \\
&= 1/2^{T/\delta t}(1 - 4a^2\delta t)^{T/(2\delta t)}\left(\left(\frac{1 + a\delta x}{1 - a\delta x}\right)^{1/\delta x}\right)^x \\
&= 1/2^{T/\delta t}exp(-a^2 T + (2a + \oslash)x + \oslash)
\end{aligned}
$$

$\square$

Notice that formula (3.14) confirms the shift of the value in the DeMoivre-Laplace approximation of $b_a(T, x)$ and $b(T, x)$. Indeed, let T be appreciable and x be limited. Let $\lambda \in \Lambda_{T,x}$. Let us write $pr(\lambda)$ its probability if it is supposed to be a trajectory of a binomial process with parameter $1/2$ and $pr^{(a)}(\lambda)$ its probability if it is supposed to be a trajectory of a binomial process with parameter p_a. Then we may derive the DeMoivre-Laplace approximation of $b_a(T, x)$ from the DeMoivre-Laplace approximation of $b(T, x)$ as follows:

$$
\begin{aligned}
b_a(T, x) &= b(T, x)\frac{pr^{(a)}(\lambda)}{pr(\lambda)} \simeq \exp-\frac{x^2}{2T} \cdot \exp(2ax - 2a^2 T) \\
&= \exp\left(-\frac{(x - 2aT)^2}{2T}\right)
\end{aligned}
$$

Proposition 3.24   *A binomial process is a Markov process.*

Proof :   Let η be a binomial process with parameter p. Let $T \in \mathbb{T}$ and suppose that η is defined on $\mathcal{C}_{[0..T]}$. For $t < T$, let (t, x) be a point of $\mathcal{C}_{[0..T]}$ and let λ be a trajectory in $\Lambda_{t,x}$. To each $\alpha \in \overline{\lambda_t^+}$, there corresponds a unique $\beta = \beta(\alpha) \in \overline{\lambda_t^-}$ such that, at any time $s < T$, $s \neq t$, the trajectory α has an upward movement at time s if and only if β has a downward

movement at time s (see Figure 3.3). By conditional independence, we
have

$$\frac{\operatorname{pr} \beta}{\operatorname{pr} \alpha} = \frac{1-p}{p},$$

and since $\overline{\lambda}_t^{\perp}$ and $\overline{\lambda}_t^-$ have the same number of elements, we have $\operatorname{Pr} \overline{\lambda}_t^- = \left(\frac{1-p}{p}\right) \cdot \operatorname{Pr} \overline{\lambda}_t^+$. Consequently,

$$
\begin{aligned}
\operatorname{Pr}_t \left\{ \delta\eta_t = \delta\eta(t,x)^+ \right\} &= \frac{\operatorname{Pr} \overline{\lambda}_t^+}{\operatorname{Pr} \overline{\lambda}_t^+ + \operatorname{Pr} \overline{\lambda}_t^-} = \frac{1}{1 + (1-p)/p} = p \\
&= \operatorname{Pr}_{t,x} \left\{ \delta\eta_t = \delta\eta(t,x)^+ \right\}.
\end{aligned}
$$

Similarly,

$$\operatorname{Pr}_t \left\{ \delta\eta_t = \delta\eta(t,x)^- \right\} = \operatorname{Pr}_{t,x} \left\{ \delta\eta_t = \delta\eta(t,x)^- \right\}.$$

Hence, the process η is a Markov process.   □

3.6   Adapted random variables

Consider a random variable defined on the trajectories of some finite stochastic process. There will be times we do not know its outcome, and times we do actually know it (at least at the ultimate time-value). For instance, assume we enter the casino early in the evening to play roulette. At that moment we do not know whether at closing time we will quit with a gain or a loss, but at closing time we know. Also, we do not know our fate at ten o'clock, but at ten o'clock and afterwards we know. The mathematical notion which models the knowledge of the outcome of a random variable is the notion of *adaptedness* to time. Within this model the random variable representing our gain or loss at ten o'clock is adapted to ten o'clock, eleven o'clock, midnight ..., but not to nine o'clock.

If we do not know the outcome of a random variable at a given time, mostly the best thing to do is to calculate its conditional expectation. In practice it is not always possible to calculate effectively the conditional expectation of a random variable: random variables are defined on a set of trajectories, and this set is usually of exponential size. However among the random variables which are adapted to some time there exists a class for which effective calculation is reasonably possible. They correspond to functions defined on the binomial cone. Because they depend on the geometrical properties of the latter, they will be called *geometrically adapted*.

Definition 3.25 Let $T \in \mathbb{T}$ and η be a stochastic process indexed by $[0 \dots T]$. Let $t \in [0 \dots T]$. A random variable g is said to be (trajectorially) *adapted* to the time t if for any trajectory λ and for any $\alpha \in \overline{\lambda}_t$

$$g(\alpha) = g(\lambda)$$

We see that a random variable which is adapted to the time t may be identified with a function $\overline{g}$ defined on the classes $\overline{\lambda}_t$ of trajectories with a common history up to time t:

$$\overline{g}(\overline{\lambda}_t) = g(\lambda).$$

So $g(\lambda)$ may be considered as a function of the $t/\delta t$ values

$$\lambda(0), \lambda(\delta t), \dots, \lambda(t)$$

taken by λ up to time t. The latter values being taken at times all less or equal to t, the outcome of the random variable may be considered *known* at time t. Clearly the random variable g, which is adapted to time t, is adapted to any time $\tau \in [t \dots T]$. Indeed, let λ be a trajectory of η and let $\alpha \in \overline{\lambda}_\tau$. Then $\alpha \in \overline{\lambda}_t$, so $g(\alpha) = g(\lambda)$. This confirms the intuition of knowledge: an outcome which is known at time t is known afterwards. Generally speaking a random variable which is adapted to time t may not be adapted to time $t - \delta t$, and a fortiori to any earlier time. For instance in the case of the process β of the example of Section 3.1 the increment $\delta\beta_1$ is adapted to the times 2 and 3, but not to the times 0 and 1. This is also in line with the model of time-dependent knowledge: throwing a coin at time 2, it is simply not known at time 0 and 1 whether it is head or tail.

Examples : We consider the Wiener walk W indexed by $[0 \dots T]$ with $T \in \mathbb{T}$.

(1) Let $t \in \mathbb{T}$. Let f be any real function. Then the random variable $f(W_t)$ is adapted to t. So the random variables $2W_t$, $expW_t$, $max(W_t, 0)$ etc. are adapted to t.

(2) The increments δW_t are adapted to $t + \delta t$, not to t.

(3) Let $\rho : [0 \ldots T] \to \mathbb{R}$ be simply a function of time. At any time t the value $\rho(t)$ may be identified with a constant random variable ρ_t. Doing so, trivially ρ_t is adapted to t.

(4) Trivially, any random variable is adapted to T.

(5) Let A be an event, let g be a random variable and let $t \in \mathbb{T}$ be arbitrary. Then the trajectorial conditional probability $Pr_t A$ is adapted to t. Also the trajectorial conditional expectation $\mathbf{E}_t g$ is adapted to t.

(6) (*Strategies*) Suppose we use the Wiener walk as a model for the gains from coin tossing, as in section 3.1. Let $[0 \ldots T]$ be some time-period. A Player P betting on tail chooses the folowing playing strategie ζ.

- $t = 0$: P bets 1\$, i.e $\eta_0 = 1$.
- $t = \delta t$: In case of gain at time 0, P bets 1\$, in case of loss at time 0, P doubles the stakes. This means that for any trajectory λ of the Wiener walk

$$\zeta_{\delta t} = \left\{ \begin{array}{ll} 1 & \delta W_0(\lambda) = 1 \\ 2 & \delta W_0(\lambda) = -1 \end{array} \right.$$

- **general case:** Like at $t = \delta t$: In case of gain at time t the bet is 1\$, in case of loss the stakes are doubled, i.e.

$$\zeta_{t+\delta t} = \left\{ \begin{array}{ll} 1 & \delta W_t(\lambda) = 1 \\ 2^k & \delta W_t(\lambda) = -1, \text{ where } k \\ & \text{is the number of losses on} \\ & \text{one row, counting back from } t \end{array} \right.$$

The sequence of random variables $(\zeta_t)_{0 \le t < T}$ represents a strategy in the common sense of the word, which means that at any time the player decides the next move entirely from the information available at that time, in casu the previous gain, or in case of loss, the number of subsequent losses. Let us prove by induction that the random variables ζ_t are adapted to t.

Clearly ζ_0, being a constant, is adapted to the time 0. Assume ζ_t is adapted to t, i.e if λ is any trajectory and $\alpha \in \overline{\lambda}_t$ we have

$$\zeta_t(\lambda) = \zeta_t(\alpha).$$

Then $\zeta_{t+\delta t}(\beta) = 1$ for all $\beta \in \overline{\lambda}_t^+$ and $\zeta_{t+\delta t}(\beta) = 2\zeta_t(\gamma)$ for all $\gamma \in \overline{\lambda}_t^-$. The trajectory λ being arbitrary, we see that $\zeta_{t+\delta t}$ is constant on any set of trajectories having a common history up to $t + \delta t$, which means that $\zeta_{t+\delta t}$ is adapted to $t + \delta t$.

Definition 3.26 Let $T \in \mathbb{T}$ and η be a stochastic process indexed by $[0 \ldots T]$. We will call a process ζ *adapted* to η if it consists of a sequence of random variables $(\zeta_t)_{0 \leq t \leq T}$ defined on the set of trajectories of η such that for every t the random variable ζ_t is adapted to t. Such a process will also be called a *derivative* of η. If not only the random variables ζ_t are adapted to t, but also the increments $\delta\zeta_t$ (for $0 \leq t < T$), the derivative ζ is called *predictable* (as a process indexed by $[0 \ldots T - \delta t]$).

Remark that the process ζ of example 6 above is a process adapted to the Wiener walk; if we let vary t the resulting sequence of random variables of examples 1,3 and 5 become also adapted processes.

If ζ is a predictable process, then the random variables $\zeta_{t+\delta t}$ are adapted t: there outcome may effectively be considered known at time t. A typical predictable process is the sequence of conditional expectations of the increments of a process; see also the notion of *trend* in the next section.

The above notion of adapted process is a special case of the notion of adapted process as defined in [16], which again may be seen as a special case of the continuous-time notion of adapted process.

The random variables of the so-called hedging process of options in Chapter 5, will be a derivative process of the geometrical Brownian motion of Chapter 4, the latter modelling the price-process of the underlying stock. This will express the fact that, in theory, the emission of an option is riskless: by adapted hedging, i.e. using financial transactions the amount of which are calculated at each stage exactly from the then available information, we will be able to cover completely the financial obligations originating from the emission.

In the context of recombining processes we turn now to so-called geometrically adapted random variables. They have properties which take

advantage from the structure of the binomial cone, facillating the calculation of their expectations.

Definition 3.27   Let $T \in \mathbb{T}$, and let η be a recombining process defined
on $\mathcal{C}_{[0..T]}$. Let Ω denote the set of all trajectories of η. A random variable
$g : \Omega \to \mathbb{R}$ is said to be *geometrically adapted to the time* $t \in \mathbb{T}$, where
$t \leq T$, if for any $x \in \mathcal{C}_t$ and for any $\lambda, \lambda' \in \Lambda_{t,x}$ we have $g(\lambda) = g(\lambda')$.

We see that a random variable g, which is geometrically adapted to the
time t may be identified with a mapping defined on the classes $\Lambda_{t,x}$, or
equivalently. to a mapping $\tilde{g}$ on the section $\mathcal{C}_t$ of the binomial cone, i.e.

$$\tilde{g}(x) = g(\lambda),$$

where λ is any trajectory of $\Lambda(t, x)$. So at time t a geometrically adapted
random variable may be considered to be a function of the state of the
underlying process at time t, and a fortiori of the actual value which is
assumed by the trajectories at time t. We will often identify the random
variable g and the mapping $\tilde{g}$. Like it was the case of trajectorially adapted
random variables the outcome of a geometrically adapted random variable
is to be considered *known* at time t. We note, however, a difference. A
random variable which is constant on $\Lambda_{t,x}$ is constant on any subclass $\overline{\lambda}_t$:
a geometrically adapted random variable is trajectorially adapted: what
is known at time t is also known at time t when forgetting the history.
Conversely, a trajectorially adapted random variable is not necessarily geometrically adapted: different histories may yield different outcomes. See
exercise 3.12 for examples.

Examples :   We consider the Wiener walk W indexed by $[0 \dots T]$, with
$T \in \mathbb{T}$.

  (1) Let $0 \leq t \leq T$. Let f be any real function. Then the random
      variable $f(W_t)$ is geometrically adapted to the time t. So the tra-
      jectorially adapted random variables $2W_t$, $expW_t$, $max(W_t, 0)$ etc.
      are also geometrically adapted to t.
  (2) Let $0 \leq t \leq T$. Let A be an event. Then the state conditional
      probability $Pr_{t,.}A$ is geometrically adapted to t. Also, if g is a
      random variable the state conditional expectation $\mathbb{E}_{t,.}g$ is geomet-
      rically adapted to t.
  (3) The random variable W_t, geometrically adapted to the time t, is
      neither geometrically adapted to $t + \delta t$ nor to the time $t - \delta t$. We

notice again a difference between geometrically adapted and trajectorially adapted random variables. Once the latter are adapted, they are adapted ever after. The first notion corresponds to a snapshot: geometrically adapted random variables are usually adapted to only one time-moment.

(4) A sequence of geometrically adapted random variables $(g_t)_{0 \leq t \leq T}$ is itself a recombining process.

In the case of a binomial process, we can calculate in a straightforward way the probability distribution and the expectation of an adapted random variable by applying Proposition 3.22. We have the following result.

Proposition 3.28 *Let $a \in \mathbb{R}$ be such that $0 < p_a < 1$. Let $T \in \mathbb{T}$ and let η be a binomial process with parameter p_a defined on $\mathcal{C}_{[0..T]}$. For each $t \in \mathbb{T}$, $t \leq T$, and for each x such that $(t, x) \in \mathcal{C}_t$, the probability distribution of a random variable g geometrically adapted to the time t is given by*

$$\Pr\left\{g = \tilde{g}(x)\right\} = b_a(t, x)\delta x$$

and its expectation is

$$Eg = \sum_{|x| \leq t/\sqrt{\delta t}} b_a(t, x)\tilde{g}(x)\delta x.$$

More in general, the state conditional expectation at some instant of a random variable adapted at the next instant has also a simple expression.

Proposition 3.29 *Let $T \in \mathbb{T}$, and let η be a recombining process indexed by $[0..T]$. Let g be a random variable adapted to the moment $t + \delta t$. Then*

$$E_{t,x}(g) = p_{t,x} \cdot \tilde{g}\left(x + \sqrt{\delta t}\right) + (1 - p_{t,x}) \cdot \tilde{g}\left(x - \sqrt{\delta t}\right).$$

Proof :

$$
\begin{aligned}
\mathrm{E}_{t,x}(g) &= \sum_{\lambda \in \Lambda_{t,x}} g(\lambda) \mathrm{pr}_{t,x}(\lambda) \\
&= \sum_{\lambda \in \Lambda_{t,x}^{+}} g(\lambda) \cdot \frac{\mathrm{pr}(\lambda)}{\mathrm{Pr}\,\Lambda_{t,x}} + \sum_{\lambda \in \Lambda_{t,x}^{-}} g(\lambda) \cdot \frac{\mathrm{pr}(\lambda)}{\mathrm{Pr}\,\Lambda_{t,x}} \\
&= \tilde{g}\left(x + \sqrt{\delta t}\right) \cdot \frac{\mathrm{Pr}\,\Lambda_{t,x}^{+}}{\mathrm{Pr}\,\Lambda_{t,x}} + \tilde{g}\left(x - \sqrt{\delta t}\right) \cdot \frac{\mathrm{Pr}\,\Lambda_{t,x}^{-}}{\mathrm{Pr}\,\Lambda_{t,x}} \\
&= p_{t,x} \cdot \tilde{g}\left(x + \sqrt{\delta t}\right) + (1 - p_{t,x}) \cdot \tilde{g}\left(x - \sqrt{\delta t}\right).
\end{aligned}
$$

$\square$

We intend to prove that in the setting of Markov processes we have, for geometrically adapted random variables, equality beween the trajectorial conditional expectation and E_t and the state conditional expectation $E(t, .)$. We note that the equality may be not verified for random variables which are not geometrically adapted; a counterexample is given by exercise 11.

To start with, we express the trajectorial conditional probability in terms of state conditional probabilities: the result is a direct consequence of proposition 3.16 and the definition of Markov processes.

Proposition 3.30 *Let $T \in \mathbb{T}$. Let η be a recombining Markov process indexed by $[0 \ldots T]$. Then for every trajectory λ and every t with $0 \leq t < T$*

$$
pr_t(\lambda) = \prod_{t \leq s < T} Pr_{s,\lambda(s)} \left\{ \delta\eta = \delta\lambda(s) \right\}
$$

For example, assume the process is binomial with parameter p. Then for every trajectory λ and time t

$$
pr_t(\lambda) = p^j (1 - p)^{\nu - j} \tag{3.15}
$$

where $\nu \equiv \nu_{T-t}$ is the number of increments between t and T and j is the number of its upward movements during this period.

As a consequence, we obtain that two trajectories with a common state and future, but possibly a different past have the same conditional probability.

Lemma 3.31 *Let $T \in \mathbb{T}$. Let η be a recombining Markov process indexed by $[0 \ldots T]$. Let $0 \leq t \leq T$ and α, β be two trajectories such that $\alpha(s) = \beta(s)$ for al s with $t \leq s \leq T$. Then*

$$pr_t(\alpha) = pr_t(\beta)$$

By definition trajectorial conditional expectations depend on the common history. As shown by the next lemma and theorem in the case of Markov processes the trajectorial conditional expectations of geometrically adapted random variables depend only on the present state. State and trajectorial conditional expectation are equal: what has happened in the past is indifferent.

Lemma 3.32 *Let $T \in \mathbb{T}$. Let η be a recombining Markov process indexed by $[0 \ldots T]$. Let g be a random variable geometrically adapted to time T. let $0 \leq t \leq T$. Let λ, μ be two trajectories such that $\lambda(t) = \mu(t)$. Then*

$$E_t g(\lambda) = E_t g(\mu).$$

Proof : Define a mapping $\phi : \overline{\lambda}_t \to \overline{\mu}_t$ by

$$\phi(\alpha) = \mu_{|[0\ldots t]} \cup \alpha_{|[t+\delta t \ldots T]}$$

Clearly ϕ is a bijection. Also, $pr_t(\alpha) = pr_t(\phi(\alpha))$ by Lemma 3.32. Because $\alpha(T) = \phi(\alpha)(T)$ and g is adapted

$$g(\alpha) = \tilde{g}(\alpha(T)) = \tilde{g}(\phi(\alpha)(T)) = g(\phi(\alpha))$$

Hence, with the correspondence $\beta = \phi(\alpha)$

$$\mathrm{E}_t g(\lambda) = \sum_{\alpha \in \overline{\lambda}(t)} g(\alpha)\mathrm{pr}_t(\alpha) = \sum_{\beta \in \overline{\mu}_t} g(\beta)\mathrm{pr}_t(\beta) = \mathrm{E}_t g(\mu).$$

$\square$

Theorem 3.33   *Let $T \in \mathbb{T}$. Let η be a recombining Markov process indexed by $[0 \ldots T]$. Let g be a random variable geometrically adapted to time T. Let $t \leq T$. Then for all $\lambda \in \Lambda$*

$$E_t g(\lambda) \;=\; E_{t,x(\lambda)} g.$$

Proof :   Put $x = x(\lambda)$. Then

$$
\begin{aligned}
\mathrm{E}_{t,x} g \;&=\; \sum_{\alpha \in \Lambda_{t,x}} g(\alpha) pr_{t,x}(\alpha) \\[2mm]
&=\; \frac{1}{Pr\Lambda_{t,x}} \sum_{\overline{\beta}_t \subset \Lambda_{t,x}} \sum_{\lambda \in \overline{\beta}_t} g(\alpha) pr(\alpha) \\[2mm]
&=\; \frac{1}{Pr\Lambda_{t,x}} \sum_{\overline{\beta}_t \subset \Lambda_{t,x}} Pr\overline{\beta}_t \sum_{\lambda \in \overline{\beta}_t} g(\alpha) pr_t(\alpha) \\[2mm]
&=\; \frac{1}{Pr\Lambda_{t,x}} \sum_{\overline{\beta}_t \subset \Lambda_{t,x}} Pr\overline{\beta}_t \mathrm{E}_t g(\beta).
\end{aligned}
$$

By Lemma 3.32 all conditional expectations $E_t g(\beta)$ are equal to $E_t g(\lambda)$, whenever $\beta \in \Lambda_{t,x}$. Also, because the classes $\overline{\beta}_t$ form a partition of $\Lambda_{t,x}$

$$\frac{1}{Pr\Lambda_{t,x}} \sum_{\overline{\beta}_t \subset \Lambda_{t,x}} Pr\overline{\beta}_t \;=\; 1.$$

We conclude that

$$\mathrm{E}_{t,x(\lambda)} g \;=\; \mathrm{E}_t g(\lambda).$$

$\square$

Consider a binomial process with parameter $p \equiv p_a$, a trajectory λ and a geometrically adapted random variable g. Put $x = x(\lambda)$. In this case the conditional expectations $\mathrm{E}_t g = \sum_{\alpha \in \overline{\lambda}_t} g(\alpha) pr_t(\alpha)$ and $\mathrm{E}_{t,x} g = \sum_{\alpha \in \Lambda_{t,x}} g(\alpha) pr_{t,x}(\alpha)$ are transformed from a "discrete path-integral" into a binomial sum: all paths which at time T are in the same state have the same conditional probability, so can be grouped together. We obtain

$$\mathrm{E}_t g(\lambda) = \mathrm{E}_{t,x}(g) = \sum_{y \le (T-t)/\sqrt{\delta t}} b_a(T-t,y) g(x+y) \delta x. \qquad (3.16)$$

See exercise 3.15 .

3.7 Martingales and trends

Constant functions are, from a certain point of view, the most elementary functions: their value at any time t is the same as their initial value at time $t = 0$. Next, we will study some particular processes which play, among general processes, the role played by the constant function among all functions. They have two important characteristics:

(1) The means of the random variables which compose these processes are constant and, consequently, equal to the initial mean
(2) If we follow a trajectory until some time s, the conditional mean from s on remains invariant.

Such processes are called *martingales*. We will state the definition of martingales for bivalent processes.

Definition 3.34 Let $T \in \mathbb{T}$, and let β be a bivalent process indexed by $[0..T]$. A derivative process η is called a *martingale* (with respect to β) if $\eta_s = \mathrm{E}_s \eta_t$ for all $s, t \in \mathbb{T}$ such that $s < t$.

A process which is a martingale with respect to itself will simply be called a martingale. The Wiener walk is an example of a martingale (with respect to itself). This fact will be established after we have introduced the notion of *trend*. The trend describes the evolution of the mean in the immediate future. The trend is for a process what the (ordinary) derivative is for a function.

Definition 3.35 Let $T \in \mathbb{T}$, and let β be a bivalent process indexed by $[0..T]$. The *trend* $D\eta$ of η is the process consisting of a sequence of random variables defined, for each t such that $t < T$, by

$$D\eta_t = \mathrm{E}_t \left(\frac{\delta \eta_t}{\delta t} \right).$$

We see that the random variables $D\eta_t$ are adapted to the time $t + \delta t$: the trend is a predictable process. Let η be a recombining Markov process. Then Theorem 3.33 and Proposition 3.29 yield an operational formula to calculate the trend.

Proposition 3.36   *Let $T \in \mathbb{T}$ and β be a recombining Markov process indexed by $[0 \dots T]$, and η be a derivative. Let $t < T$. Then for every trajectory λ of β*

$$
\begin{aligned}
D\eta_t(\lambda) &= E_{t,\lambda(t)}\left(\frac{\delta\eta_t}{\delta t}\right) \\
&= \frac{p_{t,\lambda(t)}\delta\eta(t,\lambda(t))^+ + (1 - p_{t,\lambda(t)})\delta\eta(t,\lambda(t))^-}{\delta t}
\end{aligned}
$$

Proof :   Because $\eta_{t+\delta t}$ is adapted to $t + \delta t$ and η_t is adapted to t, by Theorem 3.33 we have equality between the trajectorial conditional expectations and the state conditional expectations. Put $x = \lambda(t)$. Then we have

$$
D\eta_t(\lambda) = \frac{E_t\eta_{t+\delta t}(\lambda) - E_t\eta_t(\lambda)}{\delta t} = \frac{E_{t,x}\eta_{t+\delta t} - E_{t,x}\eta_t}{\delta t} \tag{3.17}
$$

Hence $D\eta_t(\lambda)$ is equal to $E_{t,x}\delta\eta_t/\delta t$. Applying Proposition 3.29 we see that it is also equal to

$$
\begin{aligned}
\frac{p_{t,x}\eta(t,x)^+ + (1 - p_{t,x})\eta(t,x)^- - (p_{t,x}\eta(t,x) + (1 - p_{t,x})\eta(t,x))}{\delta t} &= \\
\frac{p_{t,x}\delta\eta(t,x)^+ + (1 - p_{t,x})\delta\eta(t,x)^-}{\delta t}&.
\end{aligned}
$$

$\square$

Examples :   The Wiener walk is a process with a trend identically zero. Indeed, by Proposition 3.36 we have, for any $t \in [0 \dots T]$

$$
DW_t = \frac{\frac{1}{2}\sqrt{\delta t} + \frac{1}{2}\left(-\sqrt{\delta t}\right)}{\delta t} = 0. \tag{3.18}
$$

The trend of the process $W^{(a)}$, for some a such that $0 < p_a < 1$, is given by

$$DW_t^{(a)} = \frac{\left(\frac{1}{2} + a\sqrt{\delta t}\right)\sqrt{\delta t} + \left(\frac{1}{2} - a\sqrt{\delta t}\right)\left(-\sqrt{\delta t}\right)}{\delta t} = 2a. \qquad (3.19)$$

So, the trend of $W^{(a)}$ is limited if and only if a is limited. Note that the processes $W^{(a)}$ and $\eta_t = 2at + W_t$ have the same trend.

The processes we considered in the previous examples have constant trends. In Chapter 3.8, we will come across a process with a variable trend.

Let η be some arbitrary bivalent process, indexed by, say $[0 \dots T]$ and g be some random variable, defined in terms of the trajectories of the process. An important type of martingale is given by the sequence of all its trajectorial conditional expectations $E_t g$. To prove that this sequence is indeed a martingale we prove first some properties of the type *the mean of a family of means is again a mean*. We start with the property that the actual temporary mean of a random variable of a process is the temporary mean of the temporary mean of the random variable at the next period. Then we will generalize it to multiple time-steps.

Lemma 3.37 *Let $T \in \mathbb{T}$. Let g be a random variable on the trajectories of a bivalent process indexed by $[0..T]$. Then for each $t \in T$, $t < T$, we have $E_t E_{t+\delta t}(g) = E_t(g)$.*

Proof : Let λ be any trajectory, let λ^+ be any upper prolongation of λ at time t, and λ^- be any lower prolongation. Then

$$E_t E_{t+\delta t} g(\lambda) \;=\; \sum_{\alpha \in \overline{\lambda}_t} E_{t+\delta t} g(\alpha) pr_t(\alpha)$$

$$= \;\frac{1}{Pr\overline{\lambda}_t} \left(\sum_{\alpha \in \overline{\lambda}_t^+} E_{t+\delta t} g(\alpha) pr_t(\alpha) + \sum_{\alpha \in \overline{\lambda}_t^-} E_{t+\delta t} g(\alpha) pr_t(\alpha) \right)$$

$$= \;\frac{1}{Pr\overline{\lambda}_t} \left(E_{t+\delta t} g(\lambda^+) \cdot Pr\overline{\lambda}_t^+ + E_{t+\delta t} g(\lambda^-) \cdot Pr\overline{\lambda}_t^- \right)$$

$$= \;\frac{1}{Pr\overline{\lambda}_t} \left(\sum_{\beta \in \overline{\lambda}_t^+} g(\beta) pr(\beta) + \sum_{\gamma \in \overline{\lambda}_t^-} g(\gamma) pr(\gamma) \right)$$

$$= \;\frac{1}{Pr\overline{\lambda}_t} \sum_{\alpha \in \overline{\lambda}} g(\alpha) pr(\alpha)$$

$$= \;E_t g(\lambda).$$

$\square$

Theorem 3.38 *Let $T \in \mathbb{T}$. Let g be a random variable on the trajectories of a bivalent process indexed by $[0..T]$. Let $t, s \in T$ such that $s < t$. Then $E_s E_t(g) = E_s(g)$. As a consequence the process $E_t g$ is a martingale.*

Proof : Successive applications of Lemma 3.37 going forth and back in time from t to s yield

$$\mathrm{E}_s \mathrm{E}_t g \;=\; \mathrm{E}_s \mathrm{E}_{s+\delta t} \mathrm{E}_t g = \dots$$

$$= \;\mathrm{E}_s \mathrm{E}_{s+\delta t} \mathrm{E}_{s+2\delta t} \dots \mathrm{E}_{t-2\delta t} \mathrm{E}_{t-\delta t} \mathrm{E}_t g$$

$$= \;\mathrm{E}_s \mathrm{E}_{s+\delta t} \mathrm{E}_{s+2\delta t} \dots \mathrm{E}_{t-2\delta t} \mathrm{E}_{t-\delta t} g = \dots$$

$$= \;\mathrm{E}_s g$$

$\square$

An interesting consequence of this theorem is that, under the same hypotheses, a non-conditional expectation of a conditional expectation is the non-conditional expectation itself. That is

$$\mathrm{E}\, \mathrm{E}_t g = \mathrm{E}\, g.$$

A martingale is characterized by having zero trend. The property corresponds to the characterization of constant functions by their zero derivative.

Theorem 3.39   *Let $T \in \mathbb{T}$, and let β be a bivalent process defined on $C_{[0..T]}$. Then a derivative η is a martingale if and only if, for each $t < T$, $t \in \mathbb{T}$ we have $D\eta_t = 0$.*

Proof : If η is a martingale then

$$
\begin{aligned}
D\eta_t &= \frac{1}{\delta t} \cdot E_t \left(\eta_{t+\delta t} - \eta_t \right) \\
&= \frac{1}{\delta t} \cdot \left(E_t \eta_{t+\delta t} - E_t \eta_t \right) \\
&= \frac{1}{\delta t} \cdot \left(\eta_t - \eta_t \right) = 0.
\end{aligned}
$$

Conversely, if the trend of η is identically zero, we have for each $s, t \in T$ such that $s \leq t$

$$
\begin{aligned}
E_s \eta_t - \eta_s &= E_s \left(\eta_t - \eta_s \right) \\
&= E_s \left(\sum_{s \leq r < t} \delta \eta_r \right) \\
&= E_s \left[\sum_{s \leq r < t} \left(\frac{\delta \eta_r}{\delta t} \right) \cdot \delta t \right] \\
&= \delta t \cdot E_s \left[\sum_{s \leq r < t} E_r \left(\frac{\delta \eta_r}{\delta t} \right) \right] \\
&= \delta t \cdot E_s \left(\sum_{s \leq r < t} D\eta_r \right) = 0
\end{aligned}
$$

$\square$

From this, formula (3.18) and formula (3.19), we deduce that the Wiener walk is a martingale and that the process $W^{(a)}$ with the same increments is not a martingale if $a \neq 0$. Thus, we see that a change in the conditional probabilities of a binomial martingale can be interpreted as the introduction of a non-zero trend in that same process, and inversely.

Knowing that a given process is a martingale, we may recognize other processes to be a martingale with respect to the first on behalf of the following theorem.

Theorem 3.40 *Let $T \in \mathbb{T}$ and β be a bivalent martingale indexed by $[0 \ldots T]$. Let α be a process adapted to β. Let η_0 be a constant random variable and let for $0 \le t < T$ increments $\delta\eta_t$ be given by the random variables*

$$\delta\eta_t = \alpha_t \delta\beta_t,$$

i.e. $\eta_t = \sum_{0 \le s < t} \alpha_s \delta\beta_s$. Then η is a martingale with respect to β.

Proof : Because the random variables $\delta\beta_s$ are adapted to $s + \delta s$, all random variables $(\alpha_s \delta\beta_s)_{0 \le s < t}$ are adapted to t. Hence η_t is adapted to β_t, which implies that the process η is adapted to β. Its trend satisfies

$$D\eta_t = \mathrm{E}_t \frac{\delta\eta_t}{\delta t} = \alpha_t \mathrm{E}_t \frac{\delta\beta t}{\delta t} = \alpha_t D\beta_t = 0$$

Hence η is a martingale with respect to β. □

In the case of the Wiener walk we have a converse.

Proposition 3.41 *Let $T \in \mathbb{T}$ and η be a martingale adapted to the Wiener walk W. Then there exists a sequence of adapted random variables $(\sigma_t)_{0 \le t < T}$ such that for all $t < T$*

$$\delta\eta_t = \sigma_t \delta W_t.$$

Proof : Let λ be a trajectory of the Wiener walk. Because both η_t and $\eta_{t+\delta t}$ are adapted to $t + \delta t$ the increment $\delta\eta_t$ is adapted to $t + \delta t$. Because η is a martingale

$$0 = \mathrm{E}_t \delta\eta_t = \frac{1}{2}\overline{\delta\eta}_t(\overline{\lambda}_t^+) + \frac{1}{2}\overline{\delta\eta}_t(\overline{\lambda}_t^-)$$

So $\overline{\delta\eta}_t(\overline{\lambda}_t^-)$ and $\overline{\delta\eta}_t(\overline{\lambda}_t^-))$ are reciprocals, hence they are the same multiple, say $\overline{\sigma}_t$ of $\sqrt{\delta t}$ respectively $-\sqrt{\delta t}$. Because λ is arbitrary we have thus defined an adapted random variable α_t. □

One of the fundamental theorems of stochastics says that an adapted process can be decomposed into a martingale and a predictable process.

Theorem 3.42 (decomposition theorem) *Let $T \in \mathbb{T}$ and η be a bivalent process indexed by $[0 \dots T]$. Let ζ be a derivative of η indexed by $[0 \dots T]$. Then*

$$\zeta = \check{\zeta} + \hat{\zeta},$$

Where $\hat{\zeta}$ is a martingale with respect to η, and $\check{\zeta}$ is predictable.

Proof : For $0 \le t < T$, put

$$
\begin{aligned}
\beta_t &= \mathrm{E}_t \delta \eta_t \\
\check{\zeta}_t &= \sum_{0 \le s < t} \beta_s \\
\hat{\zeta}_0 &= 0 \\
\delta \hat{\zeta}_t &= \delta \eta_t - \beta_t
\end{aligned}
$$

Then the random variable β_t is adapted to $t + \delta_t$, and the random variable $\hat{\zeta}_t = \sum_{0 \le s < t} \delta \hat{\zeta}_t$ is adapted to t, because $\hat{\zeta}_t = \eta_t - \check{\zeta}_t$. Let $\check{\zeta}_t$ be the process consisting of the random variables $(\check{\zeta}_t)_{0 \le t < T}$ and $\hat{\zeta}$ be the process consisting of the random variables $(\hat{\zeta}_t)_{0 \le t \le T}$. Notice that $\check{\zeta}$ is predictable and that the trend of $\hat{\zeta}$ is zero, because $\mathrm{E}\,\delta \hat{\zeta}_t = \beta_t - \beta_t = 0$. Also $\zeta = \check{\zeta} + \hat{\zeta}$, so ζ is the sum of a predictable process and a martingale. $\square$

The following corollary is a consequence of the decomposition theorem and proposition 3.41.

Corollary 3.43 *Let $T \in \mathbb{T}$ and η be a process adapted to the Wiener walk. Then for every $t < T$ there are adapted random variables α_t and σ_t such that the increments of η satisfy*

$$\delta \eta_t = \alpha_t \delta t + \sigma_t \delta W_t. \tag{3.20}$$

Usually a stochastic process consists of a starting value and possibilities of taking steps going forward in time. Then, one of the main purposes of stochastics is to predict the properties of the last random variable. In case the process is adapted to the Wiener walk the increments take the form (3.20). In this context, formula (3.20) is called a *stochastic difference equation*, and we could try to solve it for (properties of) η. For an example we refer to the next chapter, where we will study equations of the form

$$\delta\xi_t = \mu\xi_t\delta t + \sigma\xi_t\delta W_t.$$

We also come across processes constructed by going back in time, i.e *backward processes*. We start with a random variable, say g defined for some future time, say T, and calculate sucessively random variables at time $T-\delta t$, $T - 2\delta t$, $T - 3\delta t$ etc. (the so-called *selffinancing strategy* of option theory in Chapter 5 is based on a construction of this type, the random variable g representing the option). Our main purpose is then to calculate its initial value (the price of the option). If the backward process happens to be a martingale, this initial value is particularly easy to find, for it is equal to the expectation Eg.

3.8   Exercises

Exercise 3.1   Consider the process β of the example of Section 3.1. Show that the increments $(\delta\beta_0, \delta\beta_{\delta t}, \delta\beta_{2\delta t})$ are independent, but that the random variables $(\beta_0, \beta_{\delta t}, \beta_{2\delta t}, \beta_{3\delta t})$ are not.

Exercise 3.2   Let $\delta t > 0$, $\delta t \simeq 0$, and consider the *Poisson random walk* ρ with independent increments, defined by $\rho_0 = 0$ and

$$\delta\rho_t = \left\{ \begin{array}{ll} 1 & \text{with probability } \delta t/2 \\ 0 & \text{with probability } 1 - \delta t \\ -1 & \text{with probability } \delta t/2 \end{array} \right.$$

(1) Sketch the set of trajectories of the process, if it is indexed by the time-interval $(0, \delta t, 2\delta t, 3\delta t)$.
(2) Consider the event "there is no change until time $t \in \mathbb{T}$", that is the event for which $\delta\rho_s = 0$ for some $0 \leq s < t$, $s \in \mathbb{T}$. Evaluate the probability of this event. At which times is the probability of this event infinitely close to 1? At which times is it infinitesimal? At which times is it neither infinitesimal nor infinitely close to 1?

Exercise 3.3   Consider the Wiener random walk indexed by $[0..T]$, where T is appreciable.

(1) Show that the trajectory $\lambda(t) = \sum_{0 \leq s < t}(-\sqrt{\delta t})^{s/\delta t}$ is S-continuous on $[0..T]$.
(2) Construct a trajectory which is infinitely close to the linear function $t \mapsto t$ on $[0..T]$.
(3) Construct a trajectory which takes an infinitely large value for some $t \in [0..T]$, $t > 0$.
(4) Construct a trajectory which is limited at any point of $[0..T]$, but is not S-continuous in each point of $[0..T]$.

Exercise 3.4   Consider the Wiener walk W indexed by the interval $[0..T]$, with T unlimited, and let $t \in \mathbb{T}$, $t \leq T$. Check that

(1) if $t \simeq 0$, the mass of w_t is included in the collection of all infinitesimals, and that there exists some $\varepsilon \simeq 0$ such that $\Pr\{|w_t| \leq \varepsilon\} \simeq 1$.
(2) if t is appreciable, the mass of w_t is included in the collection of all limited reals. Use the Normal Distribution and the fact that integrals of standard continuous functions over infinitesimal intervals are infinitesimal to show that for any $\varepsilon \simeq 0$ we have $\Pr\{|w_t| \leq \varepsilon\} \simeq 0$.
(3) What is the mass of W_t if t is unlimited?

Exercise 3.5   Let $T \in \mathbb{T}$ be appreciable, and let α and β be the stochastic processes indexed by $[0..T]$, such that $\alpha_0 = 0$, $\beta_0 = 0$, and with independent increments defined respectively by

$$\delta\alpha_t = \begin{cases} \delta t & \text{with probability } 1/2 \\ -\delta t & \text{with probability } 1/2 \end{cases}$$

and

$$\delta\beta_t = \begin{cases} \delta t^{1/3} & \text{with probability } 1/2 \\ -\delta t^{1/3} & \text{with probability } 1/2 \end{cases}$$

Use exercise 3.4 to show that if $0 \lesssim t \leq T$ for some $t \in \mathbb{T}$ then the mass of $\alpha_t \simeq 0$ is contained in the collection of infinitesimals, and that there exists some unlimited ω such that $\Pr\{|\beta_t| \leq \omega\} \simeq 0$. Conclude that the mass of β contains unlimited elements.

Exercise 3.6   Let $T \in \mathbb{T}$ be appreciable, let α be limited, and consider the process η, indexed by $[0..T]$, defined by $\eta_t = 2\alpha t + W_t$. Given any real number x, for which values of a do we have $\Pr\{\eta_t \leq x\} \simeq \Pr\{W^{(a)}(T) \leq x\}$?

Exercise 3.7   Consider the process of the example of Section 3.1 with $\delta t = 1$. Show by explicit calculation that the conditional probabilities at time $t = 2$ of the event $\{\delta \beta_2 = 1\}$ are all equal to the probability of the event $\{\delta \beta_2 = 1\}$ and that the state conditional expectations at time $t = 2$ of β_3 are all different.

Exercise 3.8   Let x and y be two random variables defined on a probability space (Ω, pr). Show that they are independent if and only if for any real numbers α and β we have

$$\Pr\{x = \alpha \mid y = \beta\} = \Pr\{x = \alpha\}$$

and

$$\Pr\{y = \beta \mid x = \alpha\} = \Pr\{y = \beta\}.$$

Exercise 3.9   Consider the binomial stochastic process indexed by a two-periods time interval, with the following equiprobable trajectories: $(1, 2, 4)$, $(1, 2, 1)$, $\left(1, \frac{1}{2}, 1\right)$ and $\left(1, \frac{1}{4}, \frac{1}{2}\right)$. Show that the increments of the random variables of this process are not independent, though they are conditionally independent.

Exercise 3.10   Let $T \in \mathbb{T}$ and η be a bivalent process indexed by $[0 \ldots T]$. Assume that for every trajectory λ and every $t \in [0 \ldots T]$

$$Pr_t \{\delta \eta_t = \delta \lambda_t\} = 1/2.$$

Show that $pr(\lambda) = (1/2)^{T/\delta t}$ for all trajectories λ.

Exercise 3.11   Consider the process β of the example of Section 3.1, for which the probabilities of the trajectories have been changed as follows

$$\mathrm{pr}\,\beta^{(8)} = \mathrm{pr}\,\beta^{(7)} = \mathrm{pr}\,\beta^{(4)} = \mathrm{pr}\,\beta^{(3)} = \mathrm{pr}\,\beta^{(2)} = \mathrm{pr}\,\beta^{(1)} = 1/8$$

$$\mathrm{pr}\,\beta^{(6)} = 1/16 \quad \text{and} \quad \mathrm{pr}\,\beta^{(5)} = 3/16.$$

(1) Check that the process is recombining.
(2) For every $(t, x) \in \mathcal{C}_{[0..3\delta t]}$, evaluate the following conditional probabilities

$$\mathrm{pr}_{t,x} \{\delta \beta_t = 1\} \quad , \quad \mathrm{pr}_{t,x} \{\delta \beta_t = 1\}$$

(3) Are the increments conditionally independent?
(4) Is the process a Markov process?

Exercise 3.12   Consider the process β of the example of Section 3.1, and consider the random variable $g : \Omega \longrightarrow \mathbb{R}$ defined by

$$g(\lambda) = \begin{cases} 1 & \text{if the number of upward movements of } \lambda \\ & \text{is larger than 2} \\ 0 & \text{otherwise} \end{cases}$$

To what times is this random variable adapted?

The same question for the following random variables:

$$h(\lambda) = \begin{cases} 1 & \text{if } \lambda \text{ has two successive upward movements} \\ 0 & \text{otherwise} \end{cases}$$

$$\phi(\lambda) = \begin{cases} 1 & \text{if } \lambda(2\delta t) \geq 2 \\ 0 & \text{otherwise} \end{cases}$$

$$\psi(\lambda) = \begin{cases} 1 & \text{if } \exists t \leq 3\delta t, \psi(t) \geq 2 \\ 0 & \text{otherwise} \end{cases}$$

Verify that none of the above random variables is geometrically adapted to any time.

In the case of the random variable h, indicate a trajectory λ and a time t such that

$$\mathrm{E}_t g(\lambda) \neq \mathrm{E}_{t,\lambda(t)} g,$$

i.e., at that moment the trajectorially conditional expectation and state conditional expectation of the random variable are different.

Exercise 3.13   (A. Fruchard) Let $T \in \mathbb{T}$ and consider the Wiener walk indexed by $[0..T]$. Put $\tau = T - t$ and define the backward process $\mathcal{M}$ indexed by $[0..T]$ as follows:

$$\mathcal{M}_\tau = W_{T-t}.$$

(1) Show that the conditional probabilities $p_{\tau,x}\left(= \mathrm{Pr}_{\tau,x}\left\{\delta\mathcal{M}_\tau = \sqrt{\delta t}\right\}\right)$ are given by

$$p_{\tau,x} = \frac{1}{1 + \dfrac{b_{\tau+\delta t,x}}{b_{\tau+\delta t,x+\delta x}}}.$$

(2) Let now $\tau \lneqq T$ and put $t = T - \tau$. Indicate the values x, with $(t,x) \in \mathcal{C}$, for which $p_{\tau,x} \simeq 1$ (one is strongly pulled up), and the

values y, with $(t, y) \in \mathcal{C}$, for which $p_{\tau, y} \simeq 0$ (one is strongly pushed down).

(3) For $\tau \nleq T$, and for limited x such that $(t, x) \in \mathcal{C}$, check that

$$p_{\tau, x} = \frac{1}{2} - \frac{x}{T - \tau} \cdot \delta x \cdot (1 + \oslash)$$

Exercise 3.14   Let $\delta t > 0$, $\delta t \simeq 0$ and η be the process indexed by $[0..T]$, with conditionally independent increments, and defined by

$$\begin{cases} \eta_0 & = & 0 \\ \delta \eta_t & = & \begin{cases} t\delta t + \sqrt{\delta t} & \text{with conditional probability } \frac{1}{2} \\ t\delta t - \sqrt{\delta t} & \text{with conditional probability } \frac{1}{2} \end{cases} \end{cases}$$

(1) What is the trend of η?
(2) Give an estimation of the mean of η_t. Express the distribution of η_t in terms of the normal distribution.
(3) Let us modify the probabilities of the increments of our process as follows

$$\begin{cases} t\delta t + \sqrt{\delta t} & \text{with probability } p_t \\ t\delta t - \sqrt{\delta t} & \text{with probability } 1 - p_t \end{cases}$$

where p_t is a number depending on the time t. For which values of p_t is the process η a martingale?

Exercise 3.15   Consider the Wiener walk indexed by $[0..T]$, and let g be a random variable geometrically adapted to the time T.

(1) What are the points $(T, z) \in \mathcal{C}_{[0..T]}$ which we can reach if we start from the point (t, x) and follow some trajectory of the Wiener walk? Let $\nu = (T - t)/\delta t$. Check that

$$\mathrm{E}_{t,x} g = \sum_{0 \leq j \leq \nu} \binom{\nu_t}{j_{t,x}} \left(\frac{1}{2}\right)^{\nu} \cdot g(x + (2j - \nu)\sqrt{\delta t}).$$

(2) Deduce formula 3.16:

$$\mathrm{E}_{t,x}(g) = \sum_{y \leq t/\sqrt{\delta t}} b(T - t, y) g(x + y) \delta y.$$

(3) Let $\delta t \simeq 0$ and T be appreciable. Suppose that the function g is of S-exponential order at infinity, and for any limited value of its domain of definition, infinitely close to a function g_0 which is standard and continuous. Show that $\mathbf{E}_{t,x}(g)$ is nearly a convolution. Actually, prove that

$$\mathbf{E}_{t,x}(g) \simeq \frac{1}{\sqrt{2\pi(T-t)}} \cdot \int_{-\infty}^{\infty} \exp\left(-\frac{y^2}{2(T-t)}\right) \cdot g_0(x+y)dy.$$

Exercise 3.16 (*stopping times*) Let $T \in \mathbb{T}$ and let the Wiener walk W be indexed by $[0\ldots T]$. Consider a random variable τ with values in $[0\ldots T]$. This time-valued random variable is called a *stopping-time* if for every time $t \in [0\ldots T]$, every trajectory α and every trajectory $\beta \in \overline{\alpha}_t$

$$\tau(\alpha) = t \Longrightarrow \tau(\beta) = t$$

(1) Define τ_1 by

$$\tau_1(\lambda) = \begin{cases} \text{Min}\,\{t \in \mathbb{T} \mid \lambda(t) \ge 100\} & \text{if } \lambda \text{ crosses the} \\ & \text{the line } x = 100 \\ T & \text{otherwise.} \end{cases}$$

Is τ_1 a stopping time?"I quit when I gain \$100".
(2) Define τ_2 by

$$\tau_2(\lambda) = \begin{cases} \text{Min}\,\{t \in \mathbb{T} \mid \lambda(t) \ge 1\} & \text{if } \lambda \text{ crosses the} \\ & \text{line } x = 100 \\ 0 & \text{otherwise.} \end{cases}$$

Is τ_2 a stopping time? "I quit unless I am sure to gain \$100".
(3) Define τ_3 by

$$\tau_3(\lambda) = \begin{cases} \text{Min}\,\{t \in \mathbb{T} \mid \lambda(t) \ge 100\} & \text{if } \lambda \text{ crosses the} \\ & \text{line } x = 100 \text{ two times} \\ T & \text{otherwise.} \end{cases}$$

Is τ_3 a stopping time? "I go on unless I will sure be losing again".
Same question for the random variable τ_4 defined by

$$\tau_4(\lambda) = \begin{cases} \text{Max}\,\{t \in \mathbb{T} \mid \lambda(t) \geq 100\} & \text{if } \lambda \text{ crosses the} \\ & \text{line } x = 100 \text{ two times} \\ T & \text{otherwise.} \end{cases}$$

"I quit when I start losing again".

(4) Define τ_5 by

$$\tau_5(\lambda) = \begin{cases} \text{Min}\,(\text{Min}\,\{t \in \mathbb{T} \mid \lambda(t) \geq 100\} + 30, T) & \text{if } \lambda \text{ crosses the} \\ & \text{the line } x = 100 \\ T & \text{otherwise.} \end{cases}$$

Is τ_1 a stopping time? "When I gain \$100 I give myself 30 minutes to do better".

Exercise 3.17   (*jump-process*). Let δt be the inverse of an unlimited integer N. Consider the process J indexed by $[0 \ldots 1]$ with $J_0 = 0$ and independent increments given by

$$\delta J_t = \begin{cases} 1 & \text{probability } \delta t \\ 0 & \text{probability } 1 - \delta t \end{cases}$$

(1) What is the trend of J?
(2) Decompose the increments of J into an increment of a predictable process and an increment of a martingale.
(3) Calculate $\text{E}\,J_1$
(4) Show that the process is recombining
(5) Determine the discrete surface $\tilde{J}$, and show that it represents a plane.
(6) Calculate approximately the probability of a constant trajectory, a trajectory with one jump, two jumps, and in general, a standard number n jumps.
(7) How many trajectories are constant, have one jump, two jumps, a standard number n jumps?
(8) Show that as long n is standard, approximately

$$Pr\,\{J_1 = n\} \simeq \frac{1}{en!}$$

Also, for all $n \leq N$

$$Pr\,\{J_1 = n\} \leq \frac{1}{n!}$$

(9) What is the mass of the random variable J_1?

(10) Let $\nu \leq N$ be unlimited. Show that there is only an infinitesimal possibility of having more than ν jumps.

Chapter 4

Stock prices

4.1 The geometric Brownian motion

Definition 4.1

Let μ, σ and ξ_0 be three real numbers, where $\xi_0, \sigma > 0$. The *discrete, geometric Brownian motion* is the stochastic process with initial value ξ_0, and with increments given by

$$\delta\xi_t = \xi_t \cdot \mu\delta t + \xi_t \cdot \sigma\delta W_t$$

We present some important properties of the discrete, geometric Brownian motion, and give a meaning to the constants μ and σ. Then we will explain why such discrete processes are good models for the evolution of stock prices in time.

First, a discrete, geometric Brownian motion is recombining: an upward movement followed by a downward movement has the same outcome as a downward movement followed by an upward movement. Indeed, the relative increment on two periods equals in both cases

$$(1 + \mu\delta t + \sigma\sqrt{\delta t})(1 + \mu\delta t - \sigma\sqrt{\delta t}) = 1 + \left(\mu - \frac{\sigma^2}{2}\right)2\delta t + \mu^2(\delta t)^2 \quad (4.1)$$

This property allows us to consider this process as a mapping defined on the binomial cone, and we get the following formulae:

$$\begin{cases} \xi(t+\delta t, x+\delta x) & = & \xi(t,x) \cdot \left(1 + \mu\delta t + \sigma\sqrt{\delta t}\right) \\ \xi(t+\delta t, x-\delta x) & = & \xi(t,x) \cdot \left(1 + \mu\delta t - \sigma\sqrt{\delta t}\right) \end{cases} \tag{4.2}$$

Moreover, the discrete, geometric Brownian motion, being adapted to the Wiener walk, is a process with equiprobable trajectories. So it is a binomial process with parameter $1/2$, and the random variable $\xi(t, \cdot)$ has the binomial distribution

$$\Pr\{\xi(t, \cdot) = \xi(t, x)\} = b(t, x)\delta x.$$

This means that at any appreciable time t and for any limited value x, we have

$$\Pr\{\xi(t, \cdot) = \xi(t, x)\} = \frac{1 + \oslash}{\sqrt{2\pi t}} \cdot \exp\left(-\frac{x^2}{2t}\right) \cdot \delta x.$$

The trend $D\xi$ of the geometric Brownian motion is given by

$$D\xi_t = \mathrm{E}_t\left(\frac{\delta\xi_t}{\delta t}\right) = \frac{\frac{1}{2}\xi_t\left(\mu\delta t + \sigma\sqrt{\delta t}\right) + \frac{1}{2}\xi_t\left(\mu\delta t - \sigma\sqrt{\delta t}\right)}{\delta t} = \mu\xi_t.$$

The *rate of return* ρ of a process is, by definition, its relative conditional trend. This means that the rate of return of the geometric Brownian motion is

$$\rho = \frac{D\xi_t}{\xi_t} = \mu.$$

We introduce now the notion of volatility, which is related to the conditional variance. It measures the vehemence of the movements of the process.

Definitions 4.2 Let η be a stochastic process. The *reduced, conditional variance* of η at time t is the expression

$$s^2(\eta_t) = \mathrm{E}_t\left(\frac{\delta\eta_t - \mathrm{E}_t\delta\eta_t}{\sqrt{\delta t}}\right)^2$$

Its square root s is called the *reduced standard deviation* of η.

The *volatility V of the process η at time t* is its relative, reduced conditional standard-deviation, *i.e.*

$$V\eta_t = \frac{s(\eta_t)}{\eta_t}$$

In the case of the geometric Brownian motion, we get

$$
\begin{aligned}
V\xi_t &= \frac{1}{\xi_t} \cdot \sqrt{\mathrm{E}_t \left(\frac{\delta\xi_t - \mu\xi_t\delta t}{\sqrt{\delta t}} \right)^2} \\
&= \frac{1}{\xi_t} \cdot \frac{\sqrt{\frac{1}{2} \cdot \left(\xi_t \sigma \sqrt{\delta t}\right)^2 + \frac{1}{2} \cdot \left(-\xi_t \sigma \sqrt{\delta t}\right)^2}}{\delta t} = \sigma
\end{aligned}
$$

We see that the discrete, geometric Brownian motion is a stochastic process with constant volatility and rate of return. So, as regards to these notions, it is the simplest stochastic process.

4.2 Modelling stock prices

Let us enumerate some properties of stock prices in the actual financial markets:

(1) Transactions are very frequent (at time-intervals of the order of a second).

(2) There is a fundamental economic hypothesis: the *weak form of market efficiency*, stating that the actual value of the stock price contains all relevant past information. Modifications of the price which could be predicted from the past are already incorporated into the actual price.

(3) Price changes are, by what we just said and apart from a trend, almost entirely due to releases of unpredictable new information. As a consequence, price-changes themselves are practically unpredictable. (See [15]). Many values are possible within a short range of time.

(4) At the exception of rare occasions, the stock price does not vary much on small time periods. Usually, the stock price at the end of a day of exchange differs from the value at the beginning of the day up to the order of one percent.

To meet the requirement that any appreciable number y can nearly be attained at short range, let us show that there exists a trajectory λ and an infinitesimal time t such that $\lambda(t) \simeq y$. We obtain such a trajectory by a "shooting method", applying a sort of approximate mean-value theorem. Take $t \simeq 0$ of the form $t = \infty \cdot \sqrt{\delta t}$. Notice that

$$1 + \mu \delta t \pm \sigma \sqrt{\delta t} = 1 \pm @\sqrt{\delta t}.$$

Because

$$(1 - @\sqrt{\delta t})^{\infty/\sqrt{\delta t}} = \exp(-@ \cdot \infty) = \oslash,$$

a trajectory with only downward movements is infinitesimal at time t. Also, a trajectory with only upward movements is unlimited at time t:

$$(1 + @\sqrt{\delta t})^{\infty/\sqrt{\delta t}} = \exp(@ \cdot \infty) = \infty.$$

So $\min\{\xi(t,x) \mid x \in \mathcal{C}_t\}$ is infinitesimal and $\max\{\xi(t,x) \mid x \in \mathcal{C}_t\}$ is unlimited. It follows from Theorem 4.5 below that $\xi(t,x)$ is S-continuous in x as least as long as x is limited. So we have

$$\xi(t, x + \delta x) \simeq \xi(t, x).$$

Combining, we see that if $\overline{x} = \max\{\xi(t,x) \mid \xi(t,x) \leq y\}$

$$\xi(t,\overline{x}) \simeq \xi(t,\overline{x} + \delta x) \simeq y.$$

(4) Under the hypotheses mentioned above, almost every trajectory of the geometric Brownian motion ξ is S-continuous. This is so because, firstly, the logarithm of a trajectory of the geometric Brownian motion is almost equal to a trajectory of a Wiener walk (this is the subject of question 1 of exercise 4.1, and secondly, by the fact that, as remarked in Section 3.2, almost all trajectories of the Wiener walk are S-continuous (see also question 5 of exercise 4.1). Exercise 4.1.5 also asks for the construction of a trajectory with a jump, and for a trajectory with vibrations.

Finally, note that the trend and conditional standard deviation in the geometric Brownian motion are proportional to the stock price. This choice was made for simplicity reasons in the beginning, but it is natural. Indeed, it should be equivalent to hold a single stock and a portfolio composed of a certain number of shares of this stock amounting to the same value. One would like, then, the trend and the standard deviation of the value of the portfolio consisting of several shares to be the sum of the trends and standard deviations of the individual shares which compose that portfolio.

It is Bachelier who, in 1900, had the very innovative idea for his time of modelling stock prices by stochastic processes. He suggested that the stock prices in a market follow trajectories of an additive Brownian motion (which we call here a Wiener walk). Nowadays one chooses to model the movements of stock prices by trajectories of a multiplicative Brownian motion, for the above-mentioned proportionality reasons on the first hand, and, on the second hand, for the fact that stocks represent limited liabilities: their prices cannot be negative. A trajectory of a Wiener walk can take negative values (which is a shortcoming of Bachelier's model), that of a geometric Brownian motion cannot.

4.3   The discrete surface of trajectories

We explained how, because of the recombining property, the trajectories of a geometric Brownian motion can be seen as a discrete surface which is a graph of a mapping defined on the binomial cone $\mathcal{C}$. Next, we show that for appreciable t and limited x this "surface" is infinitely close to an exponentially curved continous surface; also, it has exponential growth in both time and space directions.

For the reasons we stated in Section 4.2, we will suppose in the remaining part of this section that the time step δt is infinitesimal, the initial stock price ξ_0 and the volatility σ are appreciable, and that the rate of return μ is limited.

Lemma 4.3   *Let $(t,0) \in \mathcal{C}$. Then*

$$\xi(t,0) = \xi_0 \cdot e^{\left(\mu - \frac{\sigma^2}{2} + \oslash\right)t}$$

Proof :   Consider the "horizontal line" drawn in Figure (4.1). By (4.1),

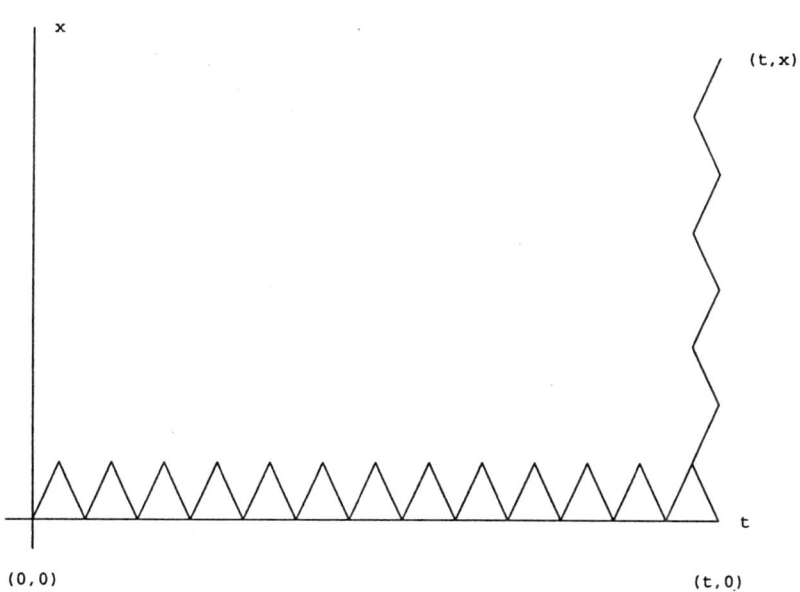

Fig. 4.1   A "horizontal line" starting from $(0,0)$ and a "vertical line" starting from $(t,0)$; here, $t = 6\delta t$

we have

$$
\begin{aligned}
\xi(t,0) &= \xi_0 \cdot \left[1 + \left(\mu - \frac{\sigma^2}{2} \right) 2\delta t + \mu^2 (\delta t)^2 \right]^{\frac{t}{2\delta t}} \\
&= \xi_0 \exp \left[\frac{t}{2\delta t} \cdot \ln \left(1 + \left(\mu - \frac{\sigma^2}{2} \right) 2\delta t + \mu^2 (\delta t)^2 \right) \right] \\
&= \xi_0 \cdot \exp \left(\frac{t}{2\delta t} \cdot (1 + \oslash) \cdot \left[\left(\mu - \frac{\sigma^2}{2} \right) 2\delta t + \mu^2 (\delta t)^2 \right] \right) \\
&= \xi_0 \cdot \exp \cdot \left(\mu - \frac{\sigma^2}{2} + \oslash \right) t
\end{aligned}
$$

$\square$

Lemma 4.4   *Let $(t,x) \in \mathcal{C}$. Then $\xi(t,x) = \xi(t,0) \exp[(1 + \oslash)\sigma x]$.*

Proof : If we follow two steps of the "vertical line" drawn in Figure (4.1), we get

$$\xi(t, x + 2\delta x) = \xi(t, x) \cdot \frac{\left(1 + \mu\delta t + \sigma\sqrt{\delta t}\right)}{\left(1 + \mu\delta t - \sigma\sqrt{\delta t}\right)}$$

$$= \xi(t, x) \cdot \frac{\left(1 + (\sigma + \oslash)\sqrt{\delta t}\right)}{\left(1 - (\sigma + \oslash)\sqrt{\delta t}\right)}$$

$$= \xi(t, x) \cdot \left(1 + (2\sigma + \oslash)\sqrt{\delta t}\right)$$

$$= \xi(t, x) \cdot (1 + (1 + \oslash)\sigma\delta x)$$

Hence, starting from the bottom of the "vertical line" we get

$$\xi(t, x) = \xi(t, 0) \cdot [1 + (1 + \oslash)\sigma\delta x]^{\left(\frac{x}{\delta x}\right)}$$

$$= \xi(t, 0) \cdot \exp\left[\frac{x}{\delta x} \cdot \ln(1 + (1 + \oslash)\sigma\delta x)\right]$$

$$= \xi(t, 0) \cdot \exp[(1 + \oslash)\sigma x]$$

$\square$

The final theorem gives an approximation of the discrete surface by means of a continuous surface. The theorem is a direct consequence of the two previous lemma's.

Theorem 4.5 *Let $(t, x) \in C$, where t and x are limited, and let $\xi(t, x)$ be a geometric Brownian motion. Then*

$$\xi(t, x) \simeq \xi_0 \cdot \exp\left[\left(\mu - \frac{\sigma^2}{2}\right)t + \sigma x\right]. \tag{4.3}$$

4.4   Exercises

Exercise 4.1 η denote the logarithm of the geometric Brownian motion ξ, indexed by $[0..T]$, where $T \in \mathbb{T}$ is unlimited. So the increments satisfy

$$\delta\eta_t = \begin{cases} \ln\left(1 + \mu\delta t + \sigma\sqrt{\delta t}\right) & \text{with conditional probability } \frac{1}{2} \\ \ln\left(1 + \mu\delta t - \sigma\sqrt{\delta t}\right) & \text{with conditional probability } \frac{1}{2} \end{cases}$$

We suppose that μ is limited, that σ is appreciable and that $\xi_0 = 1$.

(1) Let λ_t be a trajectory of η. Use a Taylor-expansion of $\delta\lambda_t$ to show that as long as t is limited, we can write $\lambda_t \simeq at + bw_t$, where w_t is a trajectory of the Wiener walk, and a and b are real numbers which may be calculated (This is a special case of an important general procedure called "Itô calculus").

(2) Estimate the trend of η. Deduce that the processes ξ and η cannot be both martingales.

(3) For $x \in \mathbb{R}$ and t appreciable, express the probability $\Pr\{\xi_t \leq x\}$ in terms of the Normal distribution. Show that $\Pr\{\xi_t \leq \epsilon\} \simeq 0$ if $\epsilon > 0$ is infinitesimal, and that $\Pr\{\xi_t \geq w\} \simeq 0$ if $w > 0$ is unlimited. What is the mass of ξ_t?

(4) Let $s \in [0 \ldots T]$. Show that

$$\mathbb{E}\,\xi_{s+\delta t} = \mathbb{E}\,\mathbb{E}\,_s\xi_{s+\delta t} = \mathbb{E}\,\xi_s(1 + \mu\delta t).$$

By iteration, obtain that for limited t

$$\mathbb{E}\,\xi_t = (1 + \mu\delta t)^{t/\delta t} \simeq \exp(\mu t).$$

(5) Show that a trajectory of the geometric Brownian motion is S-continuous for all limited t if and only if the corresponding trajectory of the Wiener walk is S-continuous for all limited t. Using, for instance exercise 3.3, exhibit a trajectory α with a jump ($\alpha(s) \not\simeq \alpha(t)$, while $s \simeq t$) and a trajectory β with vibrations, i.e. such that there are c, d with $c \not\simeq d$ and an unlimited number of times s such that $\beta(s) \simeq c$ and an unlimited number of times t such that $\beta(t) \simeq d$, all these times being infinitely close to each other.

Let us suppose in the remaining part of the exercise that $\mu = 0$, that is, ξ is a martingale.

(6) Let s be unlimited. Derive the first-order approximations $\mathbb{E}\,\eta_s = -(1 + \oslash)\sigma^2 T/2$ and $\mathrm{Var}\,\eta_s = (1 + \oslash)\sigma^2 T$. Conclude that the mass of η_s is strictly included in the negative unlimited real numbers. Show that in particular $-\ln s$ is larger than any element of the mass.

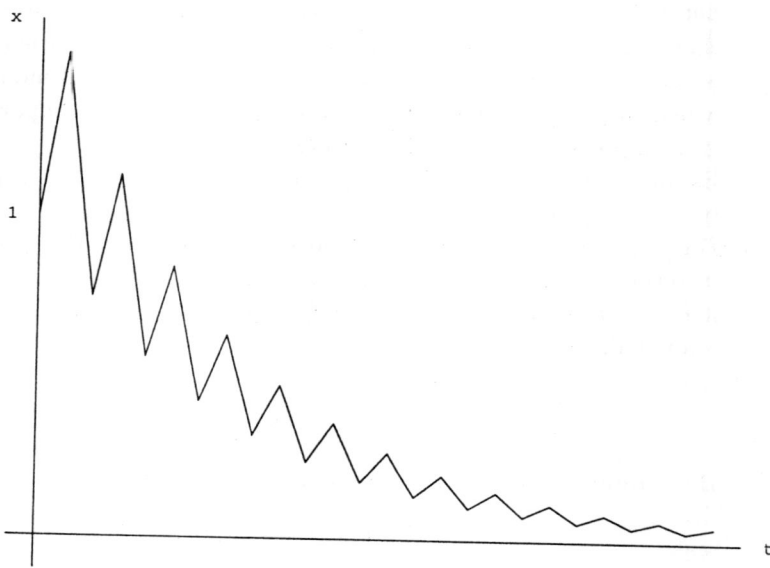

Fig. 4.2 When the geometric Brownian motion is a martingale, the "mediane" trajectory decreases.

(7) Using the previous question, show that for all unlimited $s \leq T$, we have $\Pr\{\xi_s \leq 1/s\} \simeq 1$. Deduce that in particular there exists $\varepsilon \simeq 0$ such that $\Pr\{\xi_T \leq \varepsilon\} \simeq 1$; that is, almost surely, the process leads to bankrupt even though the expectation of the random variable ξ_T is 1. (See also Figure 4.2).

(8) Consider the events $G_t = \{\xi_t \geq 1\}$ and $L_t = \{\xi_t < 1\}$. Let $E_{(t,G)}\xi_t$ and $E_{(t,L)}\xi_t$ denote the (state) conditional expectations of the random variable ξ_t with respect to these events. The first one may be seen as the expectation of the profit on the stock in the case of a gain (G) and the second as the expectation of the profit on the stock in the case of loss (L).

- Using Theorem 2.19, express, for limited t the probabilities of the events G and L and their conditional expectations in terms of the normal distribution. Then, show that

$$E_{(t,G)}\xi_t \simeq \frac{1}{E_{(t,L)}\xi_t}.$$

- Using Robinson's lemma, show that there exists an unlimited τ such that (i) for all unlimited $s \leq \tau$ we have $\Pr G \simeq 0$, (ii) $E_{(s,G)}\xi_\tau$ is unlimited, even exponentially large with respect to s, (iii) $\Pr L \simeq 1$ and finally (iv) $E_{(s,L)}\xi_\tau$ is infinitesimal, even exponentially small with respect to s.
- Formalize and derive the following property which confirms the result obtained in the previous question: in case ξ is a martingale, after a long period, a gain is almost surely a huge gain and a loss is almost surely a bankrupt.

Exercise 4.2   Consider a geometric Brownian motion ξ with appreciable initial value and limited rate of return μ. At every appreciable time τ, if the volatility is infinitesimal, each point of the mass of ξ_τ is nearly equal to ξ_0 and if the volatility is infinitely large the appreciable values are included in a set of infinitesimal probability (why?). So the volatility of a geometric Brownian motion which makes appreciable moves over an appreciable time-range with non-negligible probability (as observed on the stock-market) should be appreciable. This exercise tries to give a more intrinsic explanation of this order of magnitude.

Consider an investor observing a firm of which he possesses some stocks ξ at time t. On the whole he is rather pleased with the prize ξ_t, and thinks it matches the intrinsic value of the firm. Mathematically speaking, the value ξ_t maximizes his utility function, say U. The investor is closely following the stock-market, at regular time-intervals of infinitesimal lenght δt. He supposes his utility function to be standard, concave and twice differentiable, and to be aware of it up to a margin of $\oslash \delta t$, due to all kinds of uncertainties. Otherwise said, he observes changes in his satisfaction of the same order of magnitude as the time between two observations, i.e. of order δt.

Suppose that at time $t + \delta t$ someone offers to buy the stock at a price $\xi_{t+\delta t}$. Argue that he is only ready to sell if the price increases with at least an amount of order $\sqrt{\delta t}$. Suppose that somebody offers to sell a stock at time $t + \delta t$. Argue that he will only buy if the price drops with at least an amount of order $\sqrt{\delta t}$. Conclude that a transaction will be made only if at least $\delta\xi_t = @\sqrt{\delta t}$.

Exercise 4.3   We model the evolution of a stock price by a geometric
Brownian motion ξ. We consider a stock whose rate of return μ is limited,
but which is very volatile, that is, σ is unlimited. We suppose, however,
that $\sigma\sqrt{\delta t}$ is appreciable and suficiently small such that $\sigma\sqrt{\delta t} \lneqq 1$. We will
show that, in such conditions, the chances of going bankrupt are very large.

(1) For $t < T - \delta t$ in $\mathbb{T}$, give an approximation of $\xi(t + 2\delta t, 0)$ in terms
    of $\xi(t, 0)$, and check that $\xi(t + 2\delta t, 0) \lneqq \xi(t, 0)$.
(2) Show that $\xi(T, 0) = \exp\left(-@\sigma^2 T\right)$.
(3) Show that $\xi(T, x) = \exp\left(-@\sigma^2 T + @\sigma x\right)$ for all limited x satisfy-
    ing $(T, x) \in \mathcal{C}$. Deduce that if T is appreciable, then there exists
    $\varepsilon \simeq 0$ such that $\Pr\{\xi_T \leq \varepsilon\} \simeq 1$.

Exercise 4.4   Consider the geometric Brownian motion ξ indexed by
$[0 \ldots T]$ with appreciable T and infinitesimal time-step δt. Let the rendement μ be limited, the volatility σ be appreciable, and $\xi_0 = 1$. Let λ be
a path. We write

$$V_\lambda \equiv \sum_{0 \leq t < T} \mid \delta\lambda(t) \mid$$

the *total variation* of the path.

(1) Derive the estimation

$$V_\lambda = (1 + \oslash)\sigma \sum_{0 \leq t < T} \lambda(t)\sqrt{\delta t}$$

(2) Suppose – as is mostly the case – that the trajectory is S-continuous,
    and (see [9]) infinitely close to a standard continuous function, say
    L. Derive the estimation

$$V_\lambda = (1 + \oslash)\frac{\sigma}{\sqrt{\delta t}} \int_0^T L(t)dt$$

Let λ_1 and λ_2 be two such trajectories, with $\lambda_1(t) \leq \lambda_2(t)$ for all t
such that $0 \leq t \leq T$ and $\lambda_1(t) \lneqq \lambda_2(t)$ for all t such that $0 \lneqq t \lneqq T$
and recombining at the end, i.e. with $\lambda_1(T) = \lambda_2(T)$. Show that

$$V_{\lambda_1} \lneqq V_{\lambda_2}$$

Compare with Figure 4.2 : low values induce low variations.

(3) If λ is nearly constant, i.e., if $\lambda(t)$ is always infinitely close to $\lambda(0) = 1$, what is in first approximation the total variation of λ ? If $T \gtrsim 2$, show that there are trajectories with significantly less total variation, in fact even with limited total variation.

(4) Let us write Λ the set of trajectories and

$$V \equiv \sum_{\lambda \in \Lambda} V_\lambda (\frac{1}{2})^{T/\delta t}$$

the *mean total variation*. Below we derive a first-order approximation of V. Verify carefully the steps of this derivation.

$$
\begin{aligned}
V &= \sum_{\lambda \in \Lambda}(1+\oslash)\sigma \sum_{0 \leq t < T} \lambda(t)\sqrt{\delta t}(\frac{1}{2})^{T/\delta t} \\
&= \frac{(1+\oslash)\sigma}{\sqrt{\delta t}} \sum_{0 \leq t < T} \left(\sum_{\lambda \in \Lambda} \lambda(t)(\frac{1}{2})^{T/\delta t} \right) \delta t \\
&= \frac{(1+\oslash)\sigma}{\sqrt{\delta t}} \sum_{0 \leq t < T} e^{\mu t} \delta t \\
&= \frac{(1+\oslash)\sigma}{\sqrt{\delta t}} \int_0^T e^{\mu t}\, dt \\
&= \begin{cases} \frac{(1+\oslash)\sigma}{\mu\sqrt{\delta t}}(e^{\mu T} - 1) & \text{if } \mu \not\simeq 0 \\ \frac{(1+\oslash)\sigma T}{\sqrt{\delta t}} & \text{if } \mu \simeq 0 \,. \end{cases}
\end{aligned}
$$

Exercise 4.5 Consider the geometric Brownian motion ξ with trend μ and volatility σ. Put

$$\delta_1\xi(t,x) = \xi(t + 2\delta t, x) - \xi(t, x)$$

$$\delta_2\xi(t,x) = \xi(t, x + \delta x) - \xi(t, x)$$

$$\delta_{22}^2\xi(t,x) = \xi(t, x + 2\delta x) - 2\xi(t, x + \delta x) + \xi(t, x)$$

Suppose now that $\delta t \simeq 0$, that $\mu = 0$ and that σ is appreciable. Let t and x be any two limited real numbers such that $(t, x) \in \mathcal{C}$

(1) Show that $\frac{\delta_1\xi(t,x)}{2\delta t} \simeq -\frac{\sigma^2}{2}\xi(t,x)$ and that $\frac{\delta_2\xi(t,x)}{\delta x} \simeq \sigma\xi(t,x)$.

(2) Expand $\frac{1+\sigma\sqrt{\delta t}}{1-\sigma\sqrt{\delta t}}$ and $\left(\frac{1+\sigma\sqrt{\delta t}}{1-\sigma\sqrt{\delta t}}\right)^2$ to the second order, and check that $\frac{\delta_{22}^2\xi(t,x)}{(\delta x)^2} \simeq \sigma^2\xi(t,x)$.

(3) Deduce that, if considered as a discrete surface, the process ξ satisfies the approximative, discrete em heat equation

$$\frac{\delta_1\xi(t,x)}{2\delta t} \simeq -\frac{1}{2}\frac{\delta_{22}^2\xi(t,x)}{(\delta x)^2}.$$

N.B. For many recombining stochastic processes, the associated discrete surface satisfies approximately the heat equation. For instance, this may be shown for the option-price processes of next chapter, under some mild conditions on the option.

Chapter 5

Options

In this chapter, we consider the theory of options in the framework of finite stochastic processes. We will first introduce the fundamental notions of *option, hedge* and *self-financing strategy*. We will illustrate them by concrete examples, and then we will evaluate the price of an option. We will argue that this price takes the form of an expectation. In the recombining case the expectation can be written as a Riemann-sum. Under some conditions the Riemann-sum reduces to a Riemann-integral, representing the so-called *Black-Scholes* price.

5.1 European option, hedge, self-financing strategy

Definitions 5.1 Let $T \in \mathbb{T}$ and η be a finite stochastic process indexed by $[0..T]$. A *payoff* at time T of the random variable η_T is simply a function of the possible values of η_T. A *European option* on η *with delivery date* T is a random variable consisting of a payoff of η_T and a distribution of probabilities which is identical to that of η_T.

By the definition a European option on η with delivery date T is a random variable (trajectorially) adapted to T. Note that if η is a recombining process, a European option on η with delivery date T is a random variable geometrically adapted to the time T.

Examples :

(1) Suppose that the stock price follows a geometric Brownian motion $\xi(t, x)$. Let $T \in \mathbb{T}$ and $K > 0$. A *European call option on the stock*

ξ *with delivery date* T *and with striking price* K is, by definition, the option with the payoff

$$c(x) = (\xi(T, x) - K))^+. \tag{5.1}$$

It expresses a contract giving to its owner the right to buy the stock ξ at time T at the price K. To see this, suppose, for example, that the actual price of some stock is 110\$ and that the striking price of a European call on that stock is $K = 100\$$ with delivery date $T = 1$, let us say one year from now. After one year, there are two possibilities:

- the stock price is larger than 100\$, say $\xi(1, x) \equiv 120\$$. In this case, the owner of the call option makes use of his right to buy the stock at 100\$: he buys it and immediately sells it at the actual market price of the stock, which then equals 120\$, making a profit of 20\$ (so, in this case, the payoff of the option is 20\$). In general, if $\xi(1, x) > 100$ the profit would be exactly $\xi(1, x) - 100\$$.
- the stock price is less than or equal to 100\$. The owner of the option has the right (not the obligation) to exercise his contract. In this case, he will not make use of that right, and his profit is zero (or, otherwise stated, the payoff of the option is zero).

Combining both possibilities, we see that the payoff of the European option we considered is indeed

$$c(x) \equiv \max(\xi(1, x) - 100, 0) = (\xi(1, x) - 100)^+$$

which corresponds to formula (5.1) above.

(2) The *European put option with delivery date* T *and striking price* K is a European option with payoff at time T

$$p(x) \equiv (K - \xi(T, x))^+. \tag{5.2}$$

It is a contract giving its owner the right to sell its underlying stock ξ at time T for the price K; that is, if at time T, the stock price is less than the striking price, the owner of the option will sell the stock at a higher price, gaining the difference $K - \xi(T, x)$, and if at time T the stock price is higher than the striking price, the option will be worthless and then its payoff is zero. Summed up, the formula of the payoff of the right to sell is given indeed by (5.2).

So, a European option with respect to a process η indexed by $[0 \ldots T]$ is a random variable such that its possible values (payoff) depend only on the final value of the trajectories. If we drop the latter condition, we obtain an arbitrary random variable, the payoff being a function of the whole of the trajectories. Such an option is called *path-dependent*. An example is given by the *look-back call-option*. Its payoff is the function defined on the trajectories λ of ξ by

$$l(x) \equiv (\lambda(T) - \min_{0 \leq t \leq T} \lambda(t))^+.$$

It expresses the right to buy the stock ξ at time T for the lowest price which will occur between now and the delivery date. Another example is an Asean option. Its payoff depends on the mean of the prices during the period $[0 \ldots T]$. For instance the payoff of the Asean option which gives its owner the right to by the stock for the mean price during the period $[0 \ldots T]$ is

$$a(\lambda) \equiv \left(\lambda(T) - \frac{1}{T} \sum_{0 \leq t \leq T} \lambda(t) \delta t \right)^+.$$

But there are also more complicated financial instruments. An *American option* expresses the right to buy or sell without fixed delivery date. For example, the *American call option with striking price K and expiration date T* is a contract giving its owner the right to buy the underlying stock *at any time* $t \leq T$ for the price K, and the *American put option with striking price K and expiration date T* is a contract giving its owner the right to sell the underlying stock *at any time* $t \leq T$ for the price K. In this course, we will not consider American options. In fact, many questions concerning the valuation of such options are not answered yet: it is even not known until now how to evaluate an American put option.

Somebody who buys an option buys the right to make some very definite financial transaction at some moment. The seller, generally a broker in a financial market, the stock holder or a bank, commits himself to honour that right, whatever the stock price will be at that time. Under the assumption that the stock follows a bivalent process, the seller can follow a financial strategy which protects him *for sure* against all odds: if he follows it, he will not loose for sure! More than that, he will be sure to honour his commitment *without involving his money!* We say that such a strategy is *self-financing*. In the remaining part of this chapter, we explain how to construct and execute such a strategy, and how such a strategy determines the price of the option.

We start by introducing some concepts related to this strategy, and then we will consider some particular cases, to get acquainted with these concepts.

N.B. From now on we will only consider European or path-dependent options, and we simply refer to them as *options*.

Definitions 5.2 Let $T \in \mathbb{T}$, and let η be a finite recombining stochastic process indexed by $[0..T]$ (it will model the stock-price movements). Let v be a discrete function on $[0..T]$, which we will refer to as *a riskless investment*. A *financial strategy* is a couple of stochastic processes $(\alpha_t \eta_t, \beta_t v(t))$ where α_t and β_t are random variables which are adapted to the time t. The *value* Γ_t of the *portfolio* $(\alpha_t \eta_t, \beta_t v(t))$ is defined by

$$\Gamma_t = \alpha_t \eta_t + \beta_t v(t).$$

The sequence of values $(\Gamma_t)_{t \in [0..T]}$ is again a stochastic process. The strategy will be said to be *self-financing* if

$$\delta \Gamma_t = \alpha_t \delta \eta_t + \beta_t \delta v(t) \tag{5.3}$$

or equivalently, if

$$\Gamma_{t+\delta t} = \alpha_t \eta_{t+\delta t} + \beta_t v(t + \delta t).$$

Let f be an option. If there exists a self-financing strategy Γ such that $\Gamma_T = f$, then this strategy is called a *hedging strategy*. The initial portfolio Γ_0 is called a *hedge of the option*, and if its value is unique, we will call this value *the option price*.

In the following section, we will present some examples to illustrate these notions.

5.2 The Cox-Ross-Rubinstein model

In 1979, Cox, Ross and Rubinstein stated an explicit self-financing strategy for the replication of the price of a European option. Next, we present their method, and for a good understanding of the mechanism, we will look at the case of a European option with an expiration date which is one time-step ahead, the case of a European option with an expiration date which is two time-steps ahead, and finally the general case.

The process η modelling the price of the option's underlying stock is, as before, a geometric Brownian motion ξ. The riskless investment v of the strategy is a bond with compound interest; that is, there are two real numbers v_0 and r such that

$$v(t) = v_0(1 + r\delta t)^{\frac{t}{\delta t}}.$$

Note that α_t and β_t may be positive (buy) or negative (borrow).

5.2.1 *Case of one time-step*

Consider the case of a stock with actual price $\xi_0 = 80$. Suppose that this stock price follows a geometric Brownian motion indexed by a time interval of only two elements of length $\delta t = 1$, with rate of return $\mu = 0$ and with volatility $\sigma = \frac{1}{2}$. At the end of the period, the stock price may take two possible values:

$$\xi_1 = \begin{cases} 80 \times \left(1 + \frac{1}{2} \cdot 1\right) = 120 & \text{with probability } \frac{1}{2} \\ 80 \times \left(1 - \frac{1}{2} \cdot 1\right) = 40 & \text{with probability } \frac{1}{2} \end{cases}$$

Suppose that the initial value of a bond is $v_0 = 16$, and that the interest rate is 25%, which means that at the end of the period we have $v(1) = 20$.

Consider now a call option c giving its owner the right to buy the stock under consideration at the next time-step at its actual price; that is to say, a European option on our stock with striking price $K = 80$. The payoff of this option is then $c(\xi(1,\cdot)) = (\xi(1,\cdot) - 80)^+$, which corresponds to a random variable taking two possible values

$$\begin{cases} c(120) = 40 & \text{with probability } \frac{1}{2} \\ c(40) = 0 & \text{with probability } \frac{1}{2} \end{cases}$$

Let us now construct a self-financing strategy replicating this option. We construct a portfolio containing a fraction α of the stock and a bond of $\beta\$$.

The actual value Γ_0 of this portfolio is then $\Gamma_0 = 80\alpha + \beta$. The value of this portfolio the next moment is $\Gamma_1 = \alpha\xi_1 + \beta v(1)$, and is equal to

$$\Gamma_1 = \begin{cases} 120\alpha + 20\beta & \text{with probability } \frac{1}{2} \\ 40\alpha + 20\beta & \text{with probability } \frac{1}{2} \end{cases}$$

To replicate the option, we must have $\Gamma_1 = c$, so we have to solve the system

$$\begin{cases} 120\alpha + 20\beta & = & 40 \\ 40\alpha + 20\beta & = & 0 \end{cases}$$

The solution is $\left(\alpha = \frac{1}{2}, \beta = -1\right)$. That is to say, we can replicate the option with the portfolio $\left(\frac{1}{2}\cdot 80, -1\cdot 16\right)$. Otherwise stated, we buy half of the stock and borrow 16\$ at the interest rate of 25%. This costs $\Gamma_0 = 40 - 16 = 24$. Holding this portfolio is equivalent of holding the option c. That is why we say that the actual value of the option is $c = 24$.

Let us check that this strategy is indeed self-financing. We have

$$\begin{pmatrix} 40 \\ -16 \end{pmatrix} \longrightarrow \begin{cases} \begin{pmatrix} 60 \\ -20 \end{pmatrix} & \text{in case of an upward movement} \\ \begin{pmatrix} 20 \\ -20 \end{pmatrix} & \text{in case of a downward movement} \end{cases}$$

This strategy is trivially self-financing, because there is no extra investment of money: the value of the portfolio is due, uniquely, to the change of the value of its components. Formula (5.3) is satisfied:

$$\begin{aligned} \delta\Gamma_0 &= \Gamma_1 - \Gamma_0 \\ &= \begin{pmatrix} 40 \\ 0 \end{pmatrix} - \begin{pmatrix} 24 \\ 24 \end{pmatrix} = \begin{pmatrix} 16 \\ 16 \end{pmatrix} = \frac{1}{2}\begin{pmatrix} 40 \\ -40 \end{pmatrix} - \begin{pmatrix} 4 \\ 4 \end{pmatrix} \\ &= \alpha(\xi_1 - \xi_0) + \beta(v(1) - v(0)) = \alpha\delta\xi_0 + \beta\delta v(0) \end{aligned}$$

The example for the replication of options we just considered in the case of a unique time-step δt is fundamental. Indeed, the general case of an option with multiple time-steps will be seen as a chain of options with one time-step, and the replication of such an option will be made by a chain of self-financing strategies, each one replicating a one time-step option.

Now, we reformulate the previous example in a general context. Suppose that the initial value of the stock is $\xi_0 > 0$, and that this price is conditioned

$$P \equiv p(r) = \frac{1}{2} + \frac{r-\mu}{2\sigma}\sqrt{\delta t}. \tag{5.7}$$

(see Exercise 5.6). Summed up, in the case where the price of the stock underlying a given option follows a geometric Brownian motion, we may see the price Γ_0 of f as the actual expectation, denoted by E^r, of the option for the artificial probability $p(r)$. Hence,

$$\Gamma_0 = \frac{E^r f(\xi_{\delta t})}{1 + r\delta t}. \tag{5.8}$$

Note also that the rate of return of the backward process $(\Gamma_0, \Gamma_{\delta t})$, where

$$\Gamma_{\delta t} = \begin{cases} X & \text{with probability } p(r) \\ Y & \text{with probability } 1 - p(r) \end{cases} \tag{5.9}$$

equals the riskless rate of return r. Indeed,

$$\frac{D\Gamma(0)}{\Gamma_0} = \frac{p(r)(X - \Gamma_0) + (1 - p(r))(Y - \Gamma_0)}{\Gamma_0 \delta t} \tag{5.10}$$

$$= \frac{E^r \begin{pmatrix} X \\ Y \end{pmatrix} - \Gamma_0}{\Gamma_0 \delta t} = \frac{r\delta t \Gamma_0}{\delta t \Gamma_0} = r.$$

Notice that the rate of return must be equal to r also for economic reasons: the self-financing process being completely deterministic by construction, and covering every possible situation, it is free of risk, and *a fortiori* its rate of return must be equal to the rate of return of a riskless investment. For this reason the artificial probability $p(r)$ is often called the *risk-neutral* probability. Notice that if the probability distribution of the increments of the geometric Brownian motion ξ is changed from $(1/2, 1/2)$ into $(p(r), 1 - p(r))$, the rate of return of the process also becomes r (Exercise 5.7).

The following propositions summarize the previous reasoning.

Proposition 5.3 *Let ξ be a geometric Brownian motion on one period $(t = 0, t = \delta t)$, with non-zero volatility. Let f be a European option on ξ at time δt. Then there exists a uniquely defined, self-financing strategy replicating the option.*

Proof :   Let v be a riskless investment, and let (α, β) be the uniquely defined solution of

$$\begin{cases} \alpha\xi\left(\delta t, \sqrt{\delta t}\right) + \beta v(t) & = & f\left(\xi\left(\delta t, \sqrt{\delta t}\right)\right) \\ \alpha\xi\left(\delta t, -\sqrt{\delta t}\right) + \beta v(t) & = & f\left(\xi\left(\delta t, -\sqrt{\delta t}\right)\right) \end{cases} \tag{5.11}$$

Then the strategy

$$\left(\begin{array}{c} (\alpha\xi_0, \alpha\xi_{\delta t}) \\ (\beta v(0), \beta v(\delta t)) \end{array} \right)$$

is self-financing, and it replicates the option f. □

Proposition 5.4   *Let ξ be a geometric Brownian motion on one period $(t = 0, t = \delta t)$, with non-zero volatility σ and with rate of return μ. Suppose that the rate of interest on a riskless bond is r, and let f be a European option on ξ at the time δt. Put $p(r) = \frac{1}{2} + \frac{r-\mu}{2\sigma}\sqrt{\delta t}$. Then, the process $(\Gamma_0, \Gamma_{\delta t})$ defined by*

$$\begin{cases} \Gamma_{\delta t} & = & f(\xi_{\delta t}) \\ \Gamma_0 & = & \frac{E^r f(\xi_{\delta t})}{1+r\delta t} \end{cases} \tag{5.12}$$

is the process of values of the uniquely defined, self-financing strategy replicating the option. The rate of return of this process is r.

Proof :   By definition, we have $\Gamma_{\delta t} = f(\xi_{\delta t})$. By formulae 5.8, 5.6 and 5.11, Γ_0 is the initial value of the self-financing strategy replicating the option. By formula 5.10, the rate of return of Γ is equal to r. □

The above mentioned observations can be generalized: the price of an option is the initial value of a backward process such that its (forward) rate of return is that of a bond, and its final random variable is equal to the option. In fact, this process, if seen in the usual direction of time, will replicate the different, future possible values of the option. This observation will be made explicit in Theorem 5.5. But before stating it, we will consider the example of a process indexed by a time interval with two periods. This example is instructive because it shows the mechanism of adjusting the composition of the portfolios of the self-financing process.

5.2.2   *Case of two time-steps*

Let us reconsider the example of the previous section with a time interval consisting of two time-steps. Let $\xi = (\xi_0, \xi_1, \xi_2)$ be given by

$$\xi_0 = 80 \ , \quad \xi_1 = \begin{cases} 120 & \text{with prob. } \frac{1}{2} \\ 40 & \text{with prob. } \frac{1}{2} \end{cases} \ , \quad \xi_2 = \begin{cases} 180 & \text{with prob. } \frac{1}{4} \\ 60 & \text{with prob. } \frac{1}{2} \\ 20 & \text{with prob. } \frac{1}{4} \end{cases}$$

The riskless investment $v = (v(0), v(1), v(2))$ becomes

$$v(0) = 16 \ \longrightarrow \ v(1) = 20 \ \longrightarrow \ v(2) = 25.$$

The call option $(\xi_2 - 80)^+$ becomes

$$\begin{pmatrix} 100 \\ 0 \\ 0 \end{pmatrix} \qquad \begin{matrix} \text{with probability } \frac{1}{4} \\ \text{with probability } \frac{1}{2} \\ \text{with probability } \frac{1}{4} \end{matrix}$$

Let us construct the backward process Γ giving the values of the portfolios of the self-financing strategy which, at the last moment, replicates the call option. The initial value Γ_0 of this process will be the price of the call option under consideration. By formula (5.7), we see that the artificial probability is $p(r) = 3/4$. The backward process is then the following

$$\Gamma_0 = 36 \ , \quad \Gamma_1 = \begin{cases} 60 & \text{with prob. } \frac{3}{4} \\ 0 & \text{with prob. } \frac{1}{4} \end{cases} \ , \quad \Gamma_2 = \begin{cases} 100 & \text{with prob. } \frac{9}{16} \\ 0 & \text{with prob. } \frac{3}{8} \\ 0 & \text{with prob. } \frac{1}{16} \end{cases}$$

The different hedges can be given either by solving systems of the form of (5.4) with appropriate values, of by using directly the formulae (5.5). We find

$$\begin{cases} \alpha_0 \xi_0 & = 60 \\ \beta_0 v(0) & = -24 \end{cases} \quad \text{or} \quad \begin{cases} \alpha_0 & = \frac{3}{4} \\ \beta_0 & = -\frac{3}{2} \end{cases} \qquad (5.13)$$

$$\begin{cases} \alpha_1 \xi_1|_{\xi_1=120} & = 100 \\ \beta_1 v(1)|_{\xi_1=120} & = -40 \end{cases} \quad \text{or} \quad \begin{cases} \alpha_1|_{\xi_1=120} & = \frac{5}{6} \\ \beta_1|_{\xi_1=120} & = -2 \end{cases} \qquad (5.14)$$

$$\begin{cases} \alpha_1 \xi_1|_{\xi_1=40} & = 0 \\ \beta_1 v(1)|_{\xi_1=0} & = 0 \end{cases} \quad \text{or} \quad \begin{cases} \alpha_1|_{\xi_1=120} & = 0 \\ \beta_1|_{\xi_1=120} & = 0 \end{cases} \qquad (5.15)$$

The transactions involved in the self-financing strategy replicating the option are as follows:

t=0 Buy the portion $\alpha_0 = \frac{3}{4}$ of the stock at 80\$, amounting to 60\$, using the paid option price (36\$) and borrowing $-\beta_0 v(0) = 24\$$.

t=1 The adjustment of the portfolio depends on the stock price at this time:

If the stock price goes up then

- buy a new portion of the stock equal to $\alpha_1 - \alpha_0 = \frac{5}{6} - \frac{3}{4} = \frac{1}{12}$ by borrowing a new quantity of bonds equal to $(\beta_0 - \beta_1) v(1) = \left(-\frac{3}{2} + 2)\right) \times 20 = 10$. The amount in stocks is then, indeed, $60 \times \frac{3}{2} + 120 \times \frac{1}{12} = 100$ and the amount in bonds $-(24 \times \frac{5}{4} + 10) = -40$

If the stock price goes down then

- Sell all the stock. The proceeds of this transaction is $\alpha_0 \xi_1 = \frac{3}{4} \times 40 = 30$
- this quantity of money is equal to the debt, which at this time is equal to $-\beta_0 \cdot v(1) = -30$.

Note that there is neither gain nor loss of money in clearing the position.

t=2 There are two possible cases: $\xi_1 = 120$ or $\xi_1 = 40$.

   – If $\xi_1 = 120$, one ends up with $\frac{5}{6}$ shares of the stock and an amount equal to $-40\$$ in bonds. One should be able to close the position without investing more money. Let us check that this is the case.

If the price goes up to 180\$ then

- the option holder will ask to deliver a share for the amount of 80\$. Buy the remaining part of $\frac{1}{6}$-th of the share by borrowing 180\$ $\times \frac{1}{6} = 30\$$, then deliver the full share. Instantaneously pay back the debt, which is now $\left(-40\$ \times \frac{5}{4}\right) + (-30\$) = -80\$$, with the 80\$ paid by the option holder.

If the price goes down to 60\$ then

- close the position by selling the $\frac{5}{6}$-th of stock at the price of 60\$, which amounts to 50\$
- the proceeds of this transaction will be used to pay back the debt, which at this moment is equal to 40\$ $\times \frac{5}{4} = 50\$$

  – If $\xi_1 = 40$, there is no need of any transaction, since the position
  was already cleared in the previous period and the option will not
  be exercised.

We see that we were able to replicate any possible scenario without any
additional money. The strategy was effectively self-financing. Moreover,
the systems of equations involving ξ were not degenerated. This means
that the strategy is uniquely defined. Hence, the value $\Gamma_0 = 36$ is the
option price.

Note that there is a difference from the algebraic point of view between
the determination of the value of the portfolios of the self-financing strategy,
and the adjustment of the portfolios. The first one consists of the resolution
of a non-degenerated system of two linear equations with two variables,
which has a unique solution. The second consists of the modification of the
solutions of one linear equation with two variables, which is possible by the
non-uniqueness of the solutions.

5.2.3 *Case of multiple time-steps*

We are now able to present the fundamental theorems of option theory.
Theorem 5.5 and Theorem 5.6 state that holding an option is equivalent to
buying a portfolio and then adjusting it following a self-financing strategy.
Theorem 5.7 states that the process reproducing the price of this portfolio is
a stochastic process with conditional probabilities given by the risk-neutral
law.

Theorem 5.5   *Let ξ be a geometric Brownian motion with non-zero volatility. Let $T \in \mathrm{T}$ and let f be an option on ξ with expiration date T. Then
there exists a uniquely defined, self-financing strategy replicating $f(\xi_T)$.*

Proof :   We will explicit the strategy by working backwards. Let $v(t)$
denote the riskless investment. Suppose that the stock price at time $T' \equiv
T - \delta t$ equals $\lambda(T')$, where $\lambda(t)$ is a trajectory of ξ representing a possible
behaviour of the stock price during the period $[0 \dots T]$. Let us write λ^+ the
value at T of an upper prolongation of λ at time T' and λ^- the value at T
of a lower prolongation of λ. Consider then the "local option"

$$\begin{pmatrix} f(\lambda^+) \\ f(\lambda^-) \end{pmatrix}$$

By Proposition 5.3, there exist two uniquely defined real numbers α and β such that

$$\begin{cases} \alpha\lambda^+ + \beta v(T) &= f(\lambda^+) \\ \alpha\lambda^- + \beta v(T) &= f(\lambda^-) \end{cases} \tag{5.16}$$

The hedging quantities α and β depend only on $\lambda(T')$, and a fortiori on T', and since λ is arbitrary, generate two well-defined random variables $\alpha = \alpha_{T'}$ and $\beta = \beta_{T'}$ adapted to $T' = T - \delta t$.

Let $0 \le t < T - \delta t$. Assume, by induction, that we have defined adapted random variables α_s and β_s for all s satisfying $t < s \le T - \delta t$, and suppose that the stock price at time t is equal to $\lambda(t)$. We write as above λ^+ the value at $t + \delta t$ of an upper prolongation of λ at time t, and λ^- the value at $t + \delta t$ of a lower prolongation of λ at time t. Let α^+, β^+ be the hedging portfolio at time $t + \delta t$ for the upper prolongation, and let α^-, β^- be the hedging portfolio at time $t + \delta t$ for the lower prolongation. Consider now the local option

$$\begin{pmatrix} \alpha^+\lambda^+ + \beta^+ v(t + \delta t) \\ \alpha^-\lambda^- + \beta^- v(t + \delta t) \end{pmatrix} \tag{5.17}$$

As above, by Proposition 5.3 there exist two uniquely defined real numbers α and β such that

$$\begin{cases} \alpha\lambda^+ + \beta v(t + \delta t) = \alpha^+\lambda^+ + \beta^+ v(t + \delta t) \\ \alpha\lambda^- + \beta v(t + \delta t) = \alpha^-\lambda^- + \beta^- v(t + \delta t) \end{cases} \tag{5.18}$$

(These subtle formulae are actually the core of the finite option theory.) Since λ is arbitrary, these formulae define random variables α_t and β_t adapted to the time t. If we repeat the calculations backwards in time until $t = 0$, we will find two adapted processes α and β. By reading formulae (5.18) from the right to the left, we see that $\Gamma_{t+\delta t} = \alpha_t \xi_{t+\delta t} + \beta_t v(t+\delta t)$, hence the strategy $(\alpha_t \xi_t, \beta_t v(t))$ is self-financing and it reproduces the option $f(\xi_T)$. $\qquad\square$

We see that if we want to *determine* the self-financing strategy, we have to inverse the time and to use formulae (5.18) as equations. Inversely, to *apply* this strategy, we have to follow the natural direction of the time. Indeed, if we follow the trajectory $(\lambda(t))_{(0 \leq t \leq T)}$, we will have to buy the portfolio $(\alpha_0 \xi_0, \beta_0 v(0))$, and then to adjust it at each moment, applying the instructions given by formulae (5.18) until we come to $f(\lambda(T))$. The instructions are given at exactly the right moment: at time t we know the stock-price $\lambda(t)$ realized by the price-process ξ, hence also the values of the portfolio $(\alpha_t \lambda(t), \beta_t v(t))$, for α_t and β_t are adapted.

Note that this theorem can be immediately generalized to any bivalent process. The only property of the geometric Brownian motion which we have used is that it is non-degenerated; that is, at any moment, there are exactly two possible choices for the next movement.

In case the option is an European option, the quantities α_t and β_t are not only adapted to time, but also adapted to state. In fact the processes α and β become recombining processes, defined on the binomial triangle $\mathcal{C}_{[0...T]}$. This is proved using backward induction, along the lines of the proof of the above Theorem.

Indeed, notice first that an European option is adapted to state at time T, so it may be written in the form $f(\tilde{\xi}(T, x))$. By the induction hypothesis we may suppose that at time t the local option (5.17) has the form

$$\left(\begin{array}{c} \tilde{\alpha}(t, x)^+ \tilde{\xi}(t, x)^+ + \tilde{\beta}(t, x)^+ v(t + \delta t) \\ \tilde{\alpha}(t, x)^- \tilde{\xi}(t, x)^- + \tilde{\beta}(t, x)^- v(t + \delta t) \end{array} \right) \tag{5.19}$$

Then equation 5.18 takes the form

$$\begin{cases} \alpha \tilde{\xi}(t, x)^+ + \beta v(t + \delta t) = \tilde{\alpha}(t, x)^+ \tilde{\xi}(t, x)^+ + \tilde{\beta}(t, x)^+ v(t + \delta t) \\ \alpha \tilde{\xi}(t, x)^- + \beta v(t + \delta t) = \tilde{\alpha}(t, x)^- \tilde{\xi}(t, x)^- + \tilde{\beta}(t, x)^- v(t + \delta t) \end{cases} \tag{5.20}$$

All parameters of this system may be calculated from t and x, so the solutions may be written in the form $\alpha = \tilde{\alpha}(t, x)$ and $\beta = \tilde{\beta}(t, x)$. Since x is arbitrary, we thus obtain random variables α_t and β_t adapted to state at the time t. If we repeat the calculations backwards in time until $t = 0$, we will find two recombining processes α and β.

As a consequence, we obain that the process of values Γ is also recombining.

The above observations lead to the following theorem.

Theorem 5.6   *Let ξ be a geometric Brownian motion with non-zero volatility. Let $T \in \mathbb{T}$ and let f be a European option on ξ with expiration date T. Then there exists a uniquely defined, self-financing strategy replicating $f(\xi_T)$; the processes α and β of the hedging portfolios and the process of values Γ are recombining.*

The theorem has an obvious generalization to arbitrary recombining processes.

5.3   Black-Scholes price of European options

Notation: Consider the geometric Brownian motion with rendement μ and volatility σ. Suppose a bond has interest rate r. Then the risk-neutral probabilty equals $p(r) = \frac{1}{2} + \frac{r-\mu}{2\sigma}\sqrt{\delta t}$. We write ξ^r if ξ is supposed to be a binomial process with parameter $p(r)$. Then the probabilities of its trajectories will be written pr^r. We write E^r the expectations with respect to the probability pr^r.

Theorem 5.7   *Let ξ be a geometric Brownian motion with rate of return μ and with non-zero volatility σ. Suppose that the rate of return of a riskless investment is r. Let $T \in \mathbb{T}$ and let f be an option on ξ with expiration date T. Put $p(r) = \frac{1}{2} + \frac{r-\mu}{2\sigma}\sqrt{\delta t}$. Then the process Γ representing the price-process of the uniquely defined self-financing strategy replicating $f(\xi_T)$ is a derivative of ξ^r and its random variables satisfy*

$$\Gamma_t = \frac{1}{(1 + r\delta t)^{\frac{T-t}{\delta t}}} \cdot E_t^r f(\xi_T). \qquad (5.21)$$

Proof :   Formula (5.9) can, using the similarity of the system of equations (5.11) and the system (5.18) be generalized to any transition in the value of the replication, at any time. Suppose that at time t the value of the portfolio equals γ, with two possible transitions towards γ^+, in case the underlying trajectory of the geometric Brownian motion has an upward movement or towards γ^- in case it has a downward movement. Then we have

$$\gamma \rightarrow \begin{cases} \gamma^+ & \text{with traj. cond. prob. } p(r) \\ \gamma^- & \text{with traj. cond. prob. } 1 - p(r) \end{cases} \qquad (5.22)$$

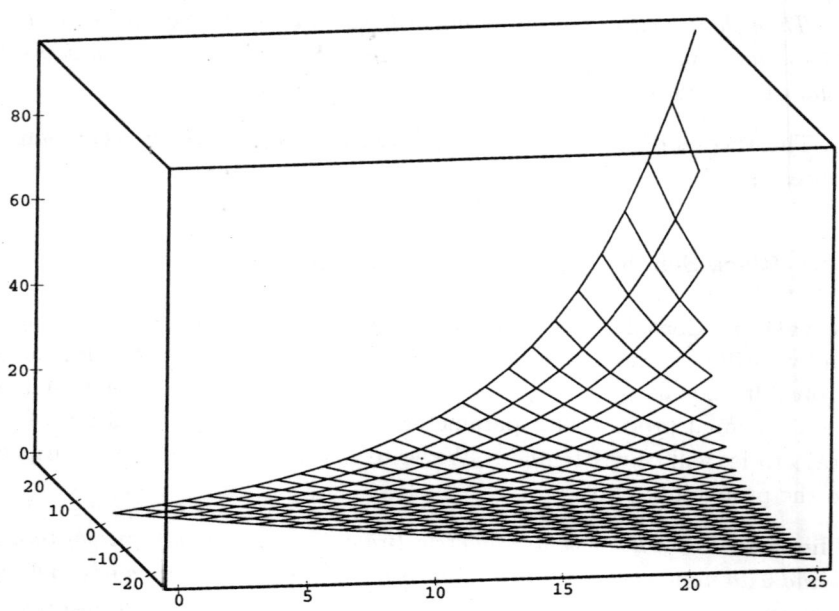

Fig. 5.1 The process of the values of the hedging portfolio's is itself a recombining process. Here, we represented it by a discrete surface in case the option is a call option.

By Proposition 3.22 the probability of a trajectory of the process Γ equals $p(r)^H.(1-p(r))^L$, where H is the number of upward movements of the underlying trajectory of ξ and L the number of its downward movements. If we attribute to this underlying trajectory the same probability we obtain ξ as a binomial process with parameter $p(r)$, so we may consider Γ as a derivative of ξ.

It follows from (5.12) that the rate of return of the process Γ is equal to r. Then, the rate of return of the process with the random variables $\frac{\Gamma_t}{(1+r\delta t)^{t/\delta t}}$ is equal to 1. So its trend is zero, which implies that the proces is a martingale. Hence, for any $t \in [0..T]$ we have

$$\frac{\Gamma_t}{(1+r\delta t)^{t/\delta t}} = \mathrm{E}_t^r \frac{\Gamma_T}{(1+r\delta t)^{T/\delta t}}.$$

Because we have identity between Γ_T and $f(\xi_T)$, we obtain formula (5.21) as a consequence. □

In case the option is European the process Γ is recombining by Theorem 5.6, in fact binomial with parameter $p(r)$, for the transitions (5.22) become

$$\tilde{\Gamma}(t,x) \;\to\; \left\{ \begin{array}{ll} \tilde{\Gamma}(t,x)^+ & \text{with state cond. prob. } p(r) \\ \tilde{\Gamma}(t,x)^- & \text{with state cond. prob. } 1-p(r) \end{array} \right.$$

The following theorem is a corollary. It presents concrete formulae for the evaluation of the price of an option.

Theorem 5.8 *Let ξ be a geometric Brownian motion with rate of return μ and with non-zero volatility σ. Suppose that the rate of return of a riskless investment is r. Let Ξ denote the set of trajectories λ of ξ. Let $T \in \mathbb{T}$ and let f be an option on ξ with expiration date T. Then, the price Γ_0 of this option is*

$$\Gamma_0 = \frac{1}{(1+r\delta t)^{\frac{T}{\delta t}}} \cdot \sum_{\lambda \in \Xi} f(\lambda) pr^r(\lambda) \tag{5.23}$$

In case the option is European the option-price reduces to

$$\Gamma_0 = \frac{1}{(1+r\delta t)^{\frac{T}{\delta t}}} \cdot \sum_{x=-T/\sqrt{\delta t}}^{T/\sqrt{\delta t}} b_{(r-\mu)/(2\sigma)}(T,x) \cdot f(\tilde{\xi}(T,x)) \cdot \delta x. \tag{5.24}$$

Proof : By Theorem 5.7, the option price is

$$\Gamma_0 = \frac{1}{(1+r\delta t)^{\frac{T}{\delta t}}} \cdot E^r f(\xi_T).$$

The formula (5.23) is a direct consequence. If the option is European, the process Γ is binomial with parameter $p(r) = \frac{1}{2} + \frac{r-\mu}{2\sigma}\sqrt{\delta t}$. Using Proposition 3.28, the expectation E^r reduces to

$$E^r f(\xi_T) = \sum_{x=-T/\sqrt{\delta t}}^{T/\sqrt{\delta t}} b_{(r-\mu)/(2\sigma)}(T,x) \cdot f(\xi(T,x)) \cdot \delta x.$$

□

A *discrete path-integral* is the mean of sums along all possible trajectories of a stochastic process. So formula (5.23) represents in a somewhat degenerated sense a discrete path-integral: for each path λ only one value $(f(\lambda))$ is taken into account. The mean is calculated for the risk-neutral probability $p(r)$.

The paths of the geometric Brownian motion are of course recombining and in case the option is European the payoff depends only of the value of the terminal values of the trajectories. This implies that the discrete path-integral (5.23) may be partitioned into sums over packets of trajectories with the same endpoint. This is what happens in formula (5.24). The weight of each packet is given by the corresponding term of the binomial distribution associated to the risk-neutral probability.

Next, we attempt to make formula (5.24) more operational. Note, first, that this formula is a Riemann sum, corresponding to an integral of the type appearing in Theorem 2.19. Moreover, in our context, this integral can be simplified because it does not depend on the risky trend μ, but on the risk-neutral trend r.

From now on, we will make the usual hypotheses on the order of magnitude of the parameters, which we recall here:

Hypotheses (H)

- the initial value ξ_0 of the discrete, geometric Brownian motion is appreciable
- the rate of return μ of the process ξ is limited
- its volatility σ^2 is appreciable
- the rate of return r of a riskless investment is limited
- the periodicity δt of transactions in time is infinitesimal
- the delivery date T of the European option is appreciable.

Theorem 5.9 (general formula for the price of a European option) *Let f be a standard, European option on the process ξ with expiration date T. Suppose the option is continuous and grows at most rationally at 0 and $+\infty$. Then, under the Hypotheses (H) above, the price Γ_0 of this option is*

$$\Gamma_0 \simeq \frac{e^{-rT}}{\sqrt{2\pi}} \int_{-\infty}^{+\infty} e^{-\frac{1}{2}y^2} \cdot f\left(\xi_0 e^{\left(r-\frac{\sigma^2}{2}\right)T + \sigma\sqrt{T}y}\right) dy \qquad (5.25)$$

Proof : Note that

$$\frac{1}{(1+r\delta t)^{T/\delta t}} \simeq e^{-rT}. \tag{5.26}$$

By Theorem 2.19 and Theorem 4.5, we have

$$\sum_{x=-T/\sqrt{\delta t}}^{T/\sqrt{\delta t}} b_{(r-\mu)/(2\sigma)}(T,x) \cdot f(\xi(T,x)) \cdot \delta x$$

$$\simeq \frac{1}{\sqrt{2\pi T}} \cdot \int_{-\infty}^{+\infty} \exp\left(\frac{-\left(x-\frac{(r-\mu)T}{\sigma}\right)^2}{2T}\right) \cdot f\left(\xi_0 e^{\left((\mu-\frac{\sigma^2}{2})T+\sigma x\right)}\right) dx.$$

We conclude by making the change of variables $y = \frac{x-(r-\mu)T/\sigma}{\sqrt{T}}$ and by applying (5.26). $\qquad\square$

Let us make some comments on the essential formula (5.25) and its derivation.

(1) The transition from the discrete formula (5.24) to the continuous formula (5.25) consists of the following chain of (near-)equalities:

$$\Gamma_0 = \frac{1}{(1+r\delta t)^{\frac{T}{\delta t}}} \cdot \sum_{x=-T/\sqrt{\delta t}}^{T/\sqrt{\delta t}} b_{(r-\mu)/(2\sigma)}(T,x) \cdot f(\tilde{\xi}(T,x)) \cdot \delta x$$

$$\simeq \frac{e^{-rT}}{\sqrt{2\pi T}} \cdot$$

$$\sum_{x=-T/\sqrt{\delta t}}^{T/\sqrt{\delta t}} \exp\left(\frac{-\left(x-\frac{(r-\mu)T}{\sigma}\right)^2}{2T}\right) \cdot f\left(\xi_0 e^{(\mu-\frac{\sigma^2}{2})T+\sigma x}\right) \delta x$$

$$\simeq \frac{e^{-rT}}{\sqrt{2\pi T}} \cdot$$

$$\int_{-\infty}^{+\infty} \exp\left(\frac{-\left(x-\frac{(r-\mu)T}{\sigma}\right)^2}{2T}\right) \cdot f\left(\xi_0 e^{(\mu-\frac{\sigma^2}{2})T+\sigma x}\right) dx$$

$$= \frac{e^{-rT}}{\sqrt{2\pi}} \cdot \int_{-\infty}^{+\infty} e^{-\frac{1}{2}y^2} \cdot f\left(\xi_0 e^{\left(r-\frac{\sigma^2}{2}\right)T+\sigma\sqrt{T}y}\right) dy.$$

Naively, in the first near-equality we replace the law of composed interest $\frac{1}{(1+r\delta t)^{T/\delta t}}$ by its Euler approximation e^{-rT}, the rescaled binomial coefficients $b_{(r-\mu)/(2\sigma)}(T,x)$ by their De Moivre-Laplace approximation $\frac{1}{\sqrt{2\pi T}}\exp\left(\frac{-\left(x-\frac{(r-\mu)T}{\sigma}\right)^2}{2T}\right)$ and the random variable of the geometric Brownian motion $\xi(T,x)$ by its continuous-surface approximation $\xi_0\exp\left((\mu-\frac{\sigma^2}{2})T+\sigma x\right)$ (formula (4.3)). This procedure is formally correct because of the exponential decay of the terms for unlimited indices.

The second near-equality transforms a summation into an integration. It is justified because the sum is the Riemann-sum of the integral; the fact that the integral is improper is immaterial, again because of the exponential decay of the tails.

The final equality is simply the result of a change of variables.

On the whole, the above transition from the discrete to the continuous is straightforward. It is based on quite common asymptotic approximations. A crucial, symplifying point is the formulation of the discrete expressions and the continuous expressions in the same coordinates (t and x): then the "continuization" consists only in replacing discrete functions by their continuous counterparts, leaving the global structure of the expressions intact.

(2) Formula (5.25) is a so-called *Feynman-Kač* formula, named after the physicist Feynman and the probabilist Kač. Such a formula gives an estimation of a *path-integral*, in this case the discrete path-integral given by formula (5.23).

(3) The rate of return μ of the stock is absent in the option-price formula (5.24). We mentioned already that it should be absent for economic reasons. We add the following mathematical argument. There is a clear relation between the trend of a process and the probabilities of its trajectories (see also Exercises 3.6 and 3.14 of Chapter 3. Now we used the geometry of the trajectories of the stock-price process to construct the self-financing process replicating the option, not their probabilities; so the latter, and *a fortiori* the risky rate of the return, should not intervene in the option price.

(4) Notice also the absence of the actual length δt of the trading periods in the price formula (5.24): the effect of the length of the trading periods is negligible, as long as it is infinitesimal.

The famous Black-Scholes formula giving the price of a European call option is now a corollary:

Theorem 5.10 (Black-Scholes formula) *Let $K > 0$ be appreciable. Put*

$$x_0 = \frac{\ln\left(\frac{\xi_0}{K}\right) + (r - \frac{\sigma^2}{2})T}{\sigma\sqrt{T}} \qquad (5.27)$$

Then, under Hypotheses (H), the price Γ_0 of a European call option with striking price K is

$$\Gamma_0 \simeq \xi_0 \mathcal{N}\left(x_0 + \sigma\sqrt{T}\right) - Ke^{-rT}\mathcal{N}(x_0). \qquad (5.28)$$

Proof : For the option $(\xi_T - K)^+$, formula (5.25) becomes

$$\Gamma_0 \simeq \frac{e^{-rT}}{\sqrt{2\pi}} \cdot \int_{-\infty}^{+\infty} e^{-\frac{1}{2}x^2} \cdot \left(\xi_0 \exp\left(\left(r - \frac{\sigma^2}{2}\right)T + \sigma\sqrt{T}x\right) - K\right)^+ dx$$

The integrand is zero until $-x_0$, which is the solution of the equation

$$\xi_0 \exp\left(\left(r - \frac{\sigma^2}{2}\right)T + \sigma\sqrt{T}x\right) = K.$$

We re-write the integral as a difference of two integrals, use the fact that the normal distribution is symmetric and put $y = \sigma T - x$ to obtain

$$\begin{aligned}
\Gamma_0 &\simeq \frac{e^{-rT}}{\sqrt{2\pi}} \cdot \int_{-x_0}^{+\infty} e^{-\frac{1}{2}x^2} \cdot \xi_0 \cdot \exp\left(rT - \frac{\sigma^2}{2}T + \sigma\sqrt{T}x\right) dx \\
&\qquad\qquad - \frac{Ke^{-rT}}{\sqrt{2\pi}} \cdot \int_{-x_0}^{+\infty} e^{-x^2/2} dx \\
&= \frac{\xi_0}{\sqrt{2\pi}} \cdot \int_{-x_0}^{+\infty} e^{-\frac{1}{2}(x - \sigma\sqrt{T})^2} dx - \frac{Ke^{-rT}}{\sqrt{2\pi}} \cdot \int_{-\infty}^{+x_0} e^{-x^2/2} dx \\
&= \frac{\xi_0}{\sqrt{2\pi}} \cdot \int_{-\infty}^{x_0 + \sigma\sqrt{T}} e^{-\frac{1}{2}y^2} dy - \frac{Ke^{-rT}}{\sqrt{2\pi}} \cdot \int_{-\infty}^{x_0} e^{-x^2/2} dx.
\end{aligned}$$

This proves formula (5.28). □

Except for being historically the first valid pricing formula to be derived, the Black-Scholes formula is notably interesting because it expresses the option-price explicitly in terms of the hedging portfolio: apparently it consists of $\mathcal{N}\left(x_0 + \sigma\sqrt{T}\right)$ stocks and and a borrowed bond amounting to

$Ke^{-rT}\mathcal{N}(x_0)$.

We have seen the method at work, and to conclude we aim to summarize it.

There are two basic models for option-theory, the finite, discrete Cox-Ross-Rubinstein model and the continuous-time Black-Scholes model. The first model has the advantage of using elementary mathematics, but the disadvantage of complicated combinatory pricing formulae, which even depend on the lenght of the time-period. The second model has the advantage of closed-form price-formulae in terms of the elementary functions of analysis, but the disadvantage of using highly advanced mathematics. The strategy of this book was to combine the two advantages, to be precise to obtain the price-formulae of the Black-Scholes model as first-order asymptotic approximations of the price-formulae of the Cox-Ross-Rubinstein model. This was possible because the *approximations* of the latter formulae do not depend on the actual lenght of the period.

Two particularities facillated the strategy. The description of the discrete model by means of a langage imitated from continuous time (we called the time variable t and the state variable x) enabled straightforward transitions from the formulae of discrete analysis to the formulae of continuous analysis; often the only essential thing to do was to replace a summation sign by an integral sign. The use of nonstandard analysis enabled us to formulate general conditions on the order of magnitude of the parameters and the regularity of the discrete functions for which the transitions are valid.

5.4   Exercises

Exercise 5.1   A "forward contract" on a stock ξ commits its owner to buy the stock at a fixed expiration date T for some fixed striking price K. Give its payoff.

Exercise 5.2   Let S be a stock. Let P be a put on S with expiration date T and exercise price K. Let C be a call on S with expiration date T for the same exercise price K. Consider now the following portfolio Π composed by a stock, a put and a borrowed call:

$$\Pi = S + P - C$$

What is the payoff of this portfolio at time T? Show that holding this portfolio is equivalent to holding a bond ("put-call parity"). What should be the actual value of Π if the rate of interest is r?

Exercise 5.3   A manager knows the contents of the report on the state of an enterprise, one month before it will be presented at the annual meeting of share-holders. The performance is surprisingly good. Which forbidden action is preferable: to buy a one-month's call with a striking price equal to the actual stock-price, or the analogous put?

Exercise 5.4   Consider a stock similar to that we considered in the course: its price ξ follows a binomial process and we have $\xi_0 = 1$ and

$$\xi \longrightarrow \begin{cases} \xi U & \text{with probability } \tfrac{1}{2} \\ \xi D & \text{with probability } \tfrac{1}{2} \end{cases}$$

Consider a European call option on the stock with striking date δt and payoff $(\xi_{\delta t} - K)^+$.

Study the sign of the fractions α and β as a function of the position of U and D with respect to K. Same question for a put option.

Exercise 5.5   (**Arbitrage:**) Consider a portfolio with actual value 0. We say that there is an *arbitrage opportunity* if at some future time its value will be positive, whatever will be the development of the stock-price, and strictly positive for at least one trajectory of the price-process.

We consider two kinds of arbitrage opportunities.

(1) Suppose that you have the possibility to buy the option of the numerical one-period example we treated in the course for 20\$. Explain how you can build a financial strategy which will provide you with a sure gain at the delivery date.
   Suppose, on the other hand, that somebody who has no knowledge of the Black-Scholes theory is ready to buy from you that same European option for 30\$. Show again how you can build a financial strategy towards a sure gain at the delivery date.

(2) Suppose that in the Cox-Ross-Rubinstein model with one period (Section 5.2.1) the interest rate R is higher than the upward movement of the stock-price U. Show that you obtain a sure gain from a well-chosen portfolio with stocks and bonds with total value 0.

The same question if the interest rate R is lower than the downward movement of the stock-price D. Describe the transactions in detail.

Exercise 5.6   We said that if the price of the stock underlying the European option follows a geometric Brownian motion, and if the riskless investment is of the form $v(t) = v_0(1 + r\delta t)$, the risk-neutral probability is $p(r) = \frac{1}{2} + \frac{r-\mu}{2\sigma}\sqrt{\delta t}$. Derive this formula, and show that, under the usual hypotheses on the orders of magnitude of the parameters, we have not only $0 < p(r) < 1$ but more precisely $p(r) = \frac{1}{2} + \pounds\sqrt{\delta t}$. Evaluate $p(r)$ in the case of the numerical example of the course.

Exercise 5.7   Show that if we replace the conditional probability p of an upward movement in the usual geometric Brownian motion by the artificial conditional probability $p(r)$, then its trend is not μ anymore, but r.

Exercise 5.8   Consider the numerical two-period option-pricing model treated in in section 5.2.2. Determine the option-price in the following cases:

(1) The option is a path-dependent call with striking price 80.
(2) The option is the Asean call $\lambda(2) - \frac{\lambda(0) + \lambda(1) + \lambda(2)}{3}$
(3) The option is an American call with striking price 80 and we choose to exercise it at the moment the payoff will be higher than the expectation of the value of the option at the next period (depending on the possible development of the stock-price).

Same questions if the options are puts.

Exercise 5.9   Assume that the possible future behaviour of the value of a given asset is described by the ordinary Wiener walk W. Let δt be the trading period, let f be an option on the asset and $T \in \mathbb{T}$ be the delivery date. Suppose the market contains also a bond v without interest.

(1) For each time $t < T$, show that there exist α_t and β_t such that $\Gamma_t \equiv \alpha_t W_t + \beta_t v(t)$ is self-financing. Express α_t and β_t in terms of $\Gamma_{t+\delta t}$. What is the value of Γ_t?
(2) Conclude that the price of the option equals $\Gamma_0 = \mathrm{E}\, f(W_T)$.

Exercise 5.10   Consider the Poisson-walk ρ of exercise 3.2 on one period $t = 0$, $t = 1$. Consider an option on η, with payoff (X,Y,Z), say. Assume a bond has interest rate $R \geq 1$.

(1) Generally speaking, is it possible to hedge the option?
(2) Suppose the market contains also a stock, modelled by a geometric Brownian motion with trading period $\delta t = 1/2$:

$$\xi \longrightarrow \begin{cases} \xi U & (U > 1) \\ \xi D & (D < 1) \end{cases}$$

Show that the option can now be hedged.

Exercise 5.11 (Cox-Ross-Rubinstein formula). Consider a geometrical Brownian motion indexed by $[0 \ldots T]$ with rate of return μ and volatility σ. Put

- $p = p(r)$,
- $D = \frac{1 + \mu \delta t - \sigma \sqrt{\delta t}}{1 + r \delta t}$,
- $U = \frac{1 + \mu \delta t + \sigma \sqrt{\delta t}}{1 + r \delta t}$ and
- $q = pU$.

Suppose that the values of the parameters are such that $0 < p < 1$ and $0 < q < 1$. Let $K > 0$. We denote by Pr_p the binomial distribution with parameter p, and by Pr_q the binomial distribution with parameter q.

(1) Show that $(1 - p)D = 1 - q$.
(2) Let $a > 0$, and let f be a random variable. By $f^{(a)}$, we will denote the random variable *truncated* to a; *i.e.* the random variable defined by

$$f^{(a)}(x) = \begin{cases} f(x) & \text{if } |f(x)| \le a \\ 0 & \text{if } |f(x)| > a \end{cases}$$

Show that $(\xi(T, x) - K)^+ = \left(\xi(T, x) - \xi(T, x)^{(K)}\right) - K \cdot \chi\{\xi(T, x) > K\}$.

(3) Show that, with $\nu \equiv \nu_T$ and $j_x \equiv j_{T,x}$,

$$
E_p \left(\frac{\xi_T - \xi_T^{(K)}}{(1 + r\delta t)^{T/\delta t}} \right)
$$

$$
= \sum_{x, \xi(T,x) > K} \xi_0 \cdot U^{j_x} \cdot D^{\nu - j_x} \cdot \binom{\nu}{j_x} \cdot p^{j_x} \cdot (1 - p)^{\nu - j_x}
$$

$$
= \xi_0 \cdot \mathrm{Pr}_q \{\xi_T > K\}.
$$

(4) Prove now the Cox-Ross-Rubinstein formula: the price Γ_0 of a call option with delivery date T and with striking price K (*i.e.* with payoff $(\xi_T - K)^+$) is equal to

$$\Gamma_0 = \xi_0 \cdot Pr_q\{\xi_T > K\} - \frac{K}{(1 + r\delta t)^{T/\delta t}} \cdot Pr_p\{\xi_T > K\}.$$

Exercise 5.12   Suppose the future evolution of a stock-price is modelled by a geometric Brownian motion ξ with rate of return μ and volatily σ. Let r be the interest rate and $T \in \mathbb{T}$ be the delivery date. Put $a \equiv (r - \mu)/(2\sigma)$, $\tau \equiv T - t$ and $g(x) \equiv f(\tilde{\xi}(T, x))$.

(1) Show that the value $\tilde{\Gamma}(t, x)$ of the hedging portfolio at time t satisfies

$$\tilde{\Gamma}(t, x) = \frac{1}{(1 + r\delta t)^{\tau/\delta t}} \sum_{|y| \leq \tau/\sqrt{\delta t}} b_a(\tau, y) g(y + x) \delta x \qquad (5.29)$$

(cf. formula 3.16).
From now on we suppose that the hypotheses (H) are satisfied, and that f is standard, continuous and of rational growth.

(2) for $t \lesssim T$ and limited x, derive the approximation

$$\tilde{\Gamma}(t, x) \simeq -\frac{e^{-r\tau}}{\sqrt{2\pi}} \int_{-\infty}^{\infty} \exp(-z^2/2) g(x + 2a\tau + \sqrt{\tau}z) dz. \qquad (5.30)$$

(3) Let x be limited. Use (5.30) and Robinson's lemma to derive that at least for some $t \simeq T$ the value of the hedging portfolio is infinitely close to the pay-off:

$$\tilde{\Gamma}(t, x) \simeq g(x) \qquad (5.31)$$

We use a different approach to show that for appreciable stock prices the value of the hedging portfolio is infinitely close to the pay-off at *all* times infinitely close to the delivery date.
To this end, we let x be limited, and we consider the binomial function $b_a(\tau, y)$ as a probability distribution on those points of the right-hand side C_T of the binomial triangle, such that their distance to x is less or equal then $\tau\sqrt{\delta t}$.

(4) Knowing the mean and standard deviation of this distribution (see exercise 1.4), show that all points of the mass of this probability distribution are infinitely close to x.

(5) Why is $g(x + y) \simeq g(x)$ for all $y \simeq 0$?

(6) Show that $b_a(\tau, y) \cdot g(y+x)$ is nearly equal to $g(x).b_a(\tau, y)$ for $y \simeq 0$, infinitesimal for limited $y \not\simeq x$ and exponentially small with respect to y for unlimited y.

(7) Derive (5.31) for all $t \simeq T$.

Exercise 5.13 Consider a geometrical Brownian motion ξ indexed by $[0 \ldots T]$ with rate of return μ and volatility σ, and a european option f on ξ. We use the notation of exercise 5.11. We know that the temporal value Γ_t of the hedging portfolio of f is a binomial process with parameter p, and that its actualized value, i.e. the process $\frac{\Gamma_t}{(1+r\delta t)}^{t/\delta t}$ is a binomial martingale.

(1) Show that the riskless component β_t of the hedging portfolio is a binomial martingale with parameter p and that its risky component α_t is a binomial martingale with parameter q.

(2) Suppose f is a call option with striking price K. What is the risky component α_T of the hedging portfolio at the delivery date? What is its riskless component β_T?

(3) Calculate now the value of α_0 and β_0. As a a consequence, derive again the formula of Cox, Ross and Rubinstein.

(4) Under the hypotheses(H), show that in first approximation

$$q = p + \frac{\sigma\sqrt{\delta t}}{2}$$

Explain now the difference $\sigma\sqrt{T}$ in the arguments of the Standard Normal distributions occurring in the Black-Scholes formula.

Exercise 5.14 Let ξ be a geometric Brownian motion with volatility σ and v be a bond, with interest rate r. Let f be a European option on ξ with expiration time $T \in \mathbb{T}$. Let α be the fraction of the stock in the hedging process, and Γ be the proces of values of the hedging portfolio. Let us consider them as discrete surfaces on the binomial cone. Let $(t, x) \in \mathcal{C}_{[0 \ldots T]}$. We use the notation for differences of exercise 4.5

(1) Show that

$$\tilde{\alpha}(t,x) = \frac{\delta_2 \tilde{\Gamma}(t,x)^-}{\delta_2 \tilde{\xi}(t,x)^-}.$$

(2) Assume that the option f is a convex function of the stock-price, as in the case of a call or a put. Then its second order difference with respect to ξ is non-negative. Using formula (5.30), show that at each time t also the second order difference of $\tilde{\Gamma}$ with respect to ξ is non-negative, implying that at each time $t < T$ the risky component $\tilde{\alpha}$ is an increasing function of ξ, hence of x. From this, and the fact, shown in exercise 5.13, that $\tilde{\alpha}(t,x)$ is a convex combination of $\tilde{\alpha}(t,x)^+$ and $\tilde{\alpha}(t,x)^-$, derive that

$$\tilde{\alpha}(t,x)^- \le \tilde{\alpha}(t,x) \le \tilde{\alpha}(t,x)^+ \tag{5.32}$$

As a consquence, show the following rule for a broker hedging a call or a put: if the price of the underlying stock increases, one should buy, and if the price decreases, one should sell. (*This self-strenghtening effect enhances the risk for leaps and bounds ...*)

(3) Assume the conditions (H) are satisfied, that f is standard continuous and of rational growth, and that $t \not< T$ and x is limited. Show that, with $a \equiv (r - \mu)/(2\sigma)$, $\tau \equiv T - t$ and $g(x) \equiv f(\tilde{\xi}(T,x))$

$$\delta_2 \tilde{\Gamma}^- = -\frac{1}{(1+r\delta t)^{(\tau-\delta t)/\delta t}} \sum_{|y| \le (\tau-\delta x)/\sqrt{\delta t}} \delta_2 b_a(\tau,y)^- g(x+y)\delta x + \pounds e^{-@/\delta x}.$$

From this, exercise 4.5 and exercise 2.15, derive that

$$\tilde{\alpha}(t,x) \simeq \frac{e^{-r\tau}}{\sigma \tilde{\xi}(t,x)\sqrt{2\pi\tau}} \int_{-\infty}^{\infty} ze^{-z^2/2} g(x+2a\tau+\sqrt{\tau}z)dz.$$

(4) Suppose f is a call option with standard striking price K. Just prior to the expiration date (at time t such that $T - t$ is of order δt), show that it is safe to hold the entire stock if it is well in the money ($\tilde{\xi}(t,x) \not> K$) and that the stock is abandoned if it is well out of the money ($\xi(t,x) \not< K$). For which values of ξ we hold a

part of the stock which is neither negligible, nor nearly total (*i.e.* such that $0 \lneq \tilde{\alpha}(t, x) \lneq 1$)?

Exercise 5.15   Suppose the stock-price process is a geometric Brownian motion ξ, and a bond v has a given rate of return r. Let f be an option on ξ with delivery date $T \in \mathbb{T}$. We use the notation ξ^r and E^r of section 5.3. In the main text, we constructed a self-financing process hedging the option, and showed (Theorem 5.7) that it consisted of a sequence of random variables, each random variable being a conditional expectation for the risk-neutral probability. Here we show a converse: the process of the sequence of risk-neutral conditional expecations is self-financing.
Indeed, consider the process

$$C_t \equiv \frac{\mathrm{E}^r_t f(\xi_T)}{(1 + r\delta t)^{(T-t)/\delta t}}.$$

(1) Show that C_t is a derivative of ξ^r.
(2) By the decomposition theorem there exists a martingale M with respect to ξ^r and a predictable process R such that at every t

$$C_t = M_t + R_t.$$

Let λ be a trajectory of ξ (or ξ^r). Verify that $\frac{\delta M_t(\gamma)}{\delta \xi_t(\gamma)}$ is constant for every trajectory $\gamma \in \overline{\lambda}_t$. Conclude that there exists an adapted random variable α_t such that $\delta M_t = \alpha_t \delta \xi_t$. Why does there exist an adapted random variable β_t such that $\delta R_t = \beta_t r \delta t$? Finally, conclude that C_t is self-financing.

Exercise 5.16   Under the hypotheses (H), evaluate the price of a European put option with appreciable striking price.

Exercise 5.17   Under the hypotheses (H), determine K (up to an infinitesimal) such that the price of a forward contract (see exercise 5.1) with striking price K is zero.

Exercise 5.18   A binary option with striking price K is an option which pays 1 in case at the delivery date T the option is "in the money", i.e. if the stock-price $\xi_T \geq K$, and pays nothing if the option is "out the money", i.e. if $\xi_T < K$. Under the hypotheses (H), what is its price?

Exercise 5.19   Consider the Black-Scholes formula for the price of a European call option with striking price K

$$\Gamma_0 \simeq \xi_0 \mathcal{N}\left(x_0 + \sigma\sqrt{T}\right) - Ke^{-rT}\mathcal{N}(x_0),\qquad(5.33)$$

where

$$x_0 = \frac{\ln\left(\frac{\xi_0}{K}\right) + (r - \frac{\sigma^2}{2})T}{\sigma\sqrt{T}}.\qquad(5.34)$$

This formula is valid for appreciable ξ_0, K, σ and T and limited r. It is both economically and mathematically interesting to see the effect when the parameters fall outside this range. We remark first that as a consequence of Robinson's lemma the Black-Scholes formula remains valid at least somewhere beyond the orders of magnitude stated above. It is supposed below that one of the parameters is infinitely small or large (but still in the range of validity of the Black-Scholes formula), while the remaining are kept appreciable or limited ("ceteris paribus").

Give a first-order approximation of the option price Γ_0, and interpret this price economically if

  (1) the striking price K is unlimited, or infinitesimal
  (2) the initial value ξ_0 is unlimited, or infinitesimal
  (3) the delivery date T is unlimited, or infinitesimal
  (4) the volatility is σ is infinitely large, or infinitesimal.

Exercise 5.20 A gentleman called Nabokov bets that at time $T \in \mathbb{T}$ a stock ξ takes exactly the – limited – value $\xi_T = \overline{\xi}$. If so he postulates an infinitely large payoff amounting to $1/\delta x$, if not, the payoff will be nothing ("butterfly option"). A banker called Dirac accepts the bet, but does not wish to involve his money. Show that under the hypotheses (H) the agreed booking price Γ_0 satisfies

$$\Gamma_0 \simeq \frac{e^{-rT}}{\sqrt{2\pi T}} \exp -\frac{\left(\ln(\overline{\xi}/\xi_0) - (r - \sigma^2/2)T\right)^2}{2\sigma^2 T}.$$

Exercise 5.21 Let ξ be a geometric Brownian motion with limited rate of return and appreciable volatily σ, and which is indexed by $[0\ldots1]$.

Consider a call-option on ξ with striking-price equal to the initial value $\xi_0 \equiv 1$. Let Γ_0 be the value of the call. Show that if the interest rate r is zero

$$\Gamma_0 \simeq \sqrt{\frac{2}{\pi}} \int_0^{\sigma/2} e^{-y^2/2} dy.$$

Conclude that the higher the volatility the higher the option price. Show that, at least up to an infinitesimal, the volatility may be determined if the option price is known. (*"Indirect volatility"*. Such a procedure is often used in practice: the volatility is not measured from a time-series, and then extrapolated to the future, but calculated from the market-prices of the option.)

Exercise 5.22 Let $T \in \mathbb{T}$ and η be a process indexed by $[0 \ldots T]$. Let Λ be the set of trajectories. An option f on η is said to be of *limited risk* if $\sum_{\lambda \in \Lambda} f(\lambda) pr(\lambda)$ is limited.
Let η be a geometric Brownian motion ξ, and suppose that the hypotheses (H) are valid.

(1) Show that a call or put with a standard striking price is of limited risk, and in general every standard continuous option with at most rational growth.
(2) Show that a butterfly option (see exercise 5.20) is of limited risk.
(3) Show that an option of limited risk has a limited price (use proposition 3.23).

Exercise 5.23 Suppose that the price of a share of some stock ξ follows a geometric Brownian motion indexed by $[0..T]$, with initial value $\xi_0 = 100$ and with volatility $\sigma = 1$. An investor speculates that the price of this stock will increase very strongly, and wants to buy it at time T for the actual price, that is $K = 100$. After studying the health of the company which released those shares, a banker came to the conclusion that the price of the stock is indeed susceptible to a very strong increase. So the banker decided to sell a call option to our investor with delivery date T, with striking price $K = 100$, but under the condition that, at the delivery date, the price of the share should not exceed $L = 150$; the contract stipulates that if the price of a share at the delivery date is larger than 150, the bank will provide the investor with no more than the amount $L - K = 50$.

(1) What is the pay-off of this "barrier" option? Express it in terms of two European call options.

(2) Knowing that the banker has access to some riskless investment with rate of return 10%, express the price of the option under consideration in terms of the normal distribution.

Exercise 5.24   (*Jump-process*) Let $\delta t \simeq 0$ and $T \in \mathbb{T}$ be appreciable. Consider the jump-process η indexed by $[0 \ldots T]$ defined by

$$\delta \eta_t \;=\; \begin{cases} \eta_t & \text{cond. prob. } \delta t \\ 0 & \text{cond. prob. } 1 - \delta t \end{cases}$$

$$\eta_0 \;>\; 0 \text{ appreciable}$$

The interest rate $r > 0$ is supposed to be limited. Consider a standard European option f on η with at most rational growth at positive infinity. Let Γ_0 be its price.

(1) Why can the option be hedged? What is the risk-neutral probability? What is the exact price formula in terms of the binomial coefficients?

(2) Using exercise 2.18, show that in first approximation

$$\Gamma_0 \simeq e^{-2rT} \sum_{j=0}^{T/\delta t} \frac{(rT)^j}{j!} f(\eta^0 \cdot 2^j)$$

(3) In case f is a call option wih standard striking-price K, derive the following price-formula in terms of the Poisson distribution $\psi(j, \lambda) \equiv \frac{\lambda^j e^{-\lambda}}{j!}$:

$$\Gamma \simeq \eta_0 \cdot \sum_{j \geq {^2}\log K} \psi(j, 2rT) - K e^{-rT} \cdot \sum_{j \geq {^2}\log K} \psi(j, rT)$$

Exercise 5.25   Suppose the stock-price follows a trajectory of a geometric Brownian motion ξ with initial value $\xi_0 = 1$, rate of return μ and volatility $\sigma = 1$. A company offers its staff the following two-term bonus: In case the stock-price at time $t = 1$ is larger than the actual price, a call-option will be presented with delivery date one year ahead, and striking price $K = 1$. In order to honour this future commitment, the financial department is held to make a reservation of a certain amount of money C. For simplicity of formulae we assume that the interest rate equals 0.

(1) Is the bonus path-dependent or path-independent?
(2) Show that C may be considered as the value of a European option Γ with delivery date 1. What is the exact pay-off of Γ? Show that it has the representation

$$\Gamma \begin{cases} = (1 + \varnothing)\xi_1 & \xi_1 \simeq +\infty \\ \simeq \xi_1 \mathcal{N}(\ln \xi_1 + \tfrac{1}{2}) - \mathcal{N}(\ln \xi_1 - \tfrac{1}{2}) & \xi_1 \geq 1 \text{ limited} \\ = 0 & \xi_1 \leq 1. \end{cases}$$

(3) Express C in terms of the normal distribution.
(4) Suppose the bonus is offered in the following form: at $t = 1$ independently of the stock-price a call-option wil be presented with delivery date one year ahead and striking price 1. A bright employee sees immediately what reservation should be made by the financial department. How much is this reservation?

Bibliography

1. Anderson, R.M. (1976) ''A Nonstandard Representation for Brownian Motion and Itô Integration'', *Israel Math. J.*, **25**, 15.
2. Berg I.P. van den (1987) *Nonstandard Asymptotic Analysis*, Springer Verlag, Berlin.
3. Berg, I.P. van den and Koudjeti, F. (1997) ''From binomial expectations to the Black-Scholes formula: the main ideas'', *Ann. Math. Blaise Pascal*, **4**, 93-101.
4. Black, F. and Scholes, M. (1973) ''The pricing of Options and Corporate Liabilities''*Journal of Political Economy*, **81**, 635.
5. Cox, J.C., Ross, S.A. and Rubinstein, M. (1979) ''Option pricing, a simplified approach''*Financial Economics*, **7**, 637.
6. N.Cutland, E.Kopp, and W.Willinger (1991) ''A nonstandard approach to option pricing'', *Mathematical Finance*, **9** (16), 1.
7. Deledicq, A. and Diener, M. (1989) *Leçons de calcul infinitésimal*, Armand Colin, Paris.
8. Diener, F. and Diener, M., editors (1995) *Nonstandard Analysis in Practice*, Springer Verlag, Berlin.
9. Diener, F. and Reeb, G. (1989) *Analyse Non Standard*, Hermann, Paris.
10. Duffie, D. (1988) *Security Markets, Stochastic Models*, Academic Press, New York.
11. Feller, W. (1968) *An Introduction to Probability Theory and Its Applications*, John Wiley & sons, Inc., New York.
12. Keisler, H. J. (1976) *Elementary Calculus*, Prindle, Weber and Schmidt.
13. Koudjeti, F. (1995) *Elements of external calculus, with an application to mathematical finance*, Labyrint Publication, Capelle a/d IJssel, The Netherlands.
14. Lamberton, D. and Lapeyre, B. (1993) *Introduction au calcul stochastique appliqué à la finance*, Ellipses, Paris.
15. Malkiel, R.G. (1990) *A random walk down Wall Street*, Norton, New York.

16.Nelson, E. (1987) *Radically elementary probability theory,* Princeton University Press, Princeton.

17.Petry, A. (1996) *Analyse infinitésimale*, Imprisil, Liege, Belgium.

18.Robert, A. (1988) *Nonstandard Analysis*, John Wiley and Sons Inc. New York.

Index

134 *Index*